MW01515420

Heaven

and

Earth

and in

Between

Gail Adams

PublishAmerica
Baltimore

First printing

ISBN: 1-4137-6335-9
PUBLISHED BY PUBLISHAMERICA, LLLP
www.publishamerica.com
Baltimore

Printed in the United States of America

To God the Father
Glory to His Name

Acknowledgment

I wish to thank my brother Grant Holditch and his wife Marilyn for their prayers and guidance throughout the years.

I also wish to thank my parents Mr. and Mrs. Gilman Holditch (now deceased) for their prayers continually for my salvation. Even when things must have looked bleak for them, they never gave up. To them I thank for once I was lost and now am found.

I must thank our Lord and Saviour, for blessing me as He has.

Foreword

A born again Christian, I was raised in a Christian home, I drifted in and out of a close relationship with God. I have walked far from Him, and have wasted many good years, which I regret.

In the latter part of 2003, I asked God for the wisdom of David and Solomon, the patience of Job, and the love of Jesus. In the beginning of 2004, God began granting me my prayers. Daily He needs to work on me, for I see I take a step forward, and several back.

The day before I was notified that my manuscript was accepted for publishing, I received this poem.

Vessel

Ye are a vessel
A vessel of My choosing
I will use you for My purposes
To speak out My words
I will use you
And you will speak My words
I will use you to bring about My purposes
A vessel you are
I will use any vessel
I have used a donkey
For My purposes
Don't think you are anything special
You are a messenger
A messenger only
You retain nothing more with Me
You must come too as all must come
For I will use who I please
And do not think you have gained a step closer
Because I have used you so
For ye are nothing but a mere vessel
A vessel for My purposes
An earthen vessel
Made of clay
Full of cracks
Which leak out My word
Oh, ye are weak
Ye weakened vessel
Don't allow pride to step in
For ye are nothing
But a mere vessel to Me
To be used as I will
Ye must come to Me
As all must come
Humble before your God

Humble before your Redeemer
For I alone am God
I alone bring all things about
And I will use any vessel I choose
Don't allow pride to take you
For it desires you
For I will bring a stone to bring My purposes about
I say

Sept. 1, 2004

I believe that the writings in this book are that of the leading of the Holy Spirit, and let me never forget where I have come from, a sinner, and that I am a mere vessel. I must always remember that I will still fall into sin, and that no one is sinless, no, not one. But we are forgiven!

A Child

Oh, to watch a child run to their father
Oh, how they run
Oh, they welcome him always
Oh, how they crawl upon his knee
Telling him of their day
They cry to him when they encounter fear
They run for hugs when hurt
Oh, they enjoy just to be with their daddy
Walk with him, hand in hand
Oh, the innocence of a child
Telling him all
Nothing held back
Oh, their only fear is when they have done wrong
But then, they know a scolding perhaps, would be about to come
Then the hug, a caress, and told not to do it again
Oh, what love a child has with their daddy
Oh, how that love grows as we mature
Where we are held responsible for our own actions
But, always knowing he is there to run to when in trouble
He is ever there to bail us out
Ever there for us to just confide in
Ever there to hug us
Always there loving us
Even when we don't deserve it
Oh, to know the love of a child with their daddy
Oh, to know that
How blessed we are to know the love between a father and a child
Oh, to know that
He is always there
Always
Oh, to know that

Mar. 24 2004

A Seed Planted

I plant a seed
The seed if not planted—it will die
But if that seed is planted—it is then bonded to the earth
It grows
It sends out its seeds for the wind to take
The wind accepts these seeds
These seeds are scattered
They bond with the wind
They find good soil
They root themselves
Many—not all—find good soil
They grow—both to give off their seed to continue
But many that are lost—find themselves decayed
Like the earth they dwell in
They become like the death that devours them
But the seed that finds good earth
They are fed
Light reaches their inner being
They become life itself
They grow—they give off life to enable a tenfold of themselves reproduced
So be it

Feb. 4, 2004

All Things Right

Oh, Lord, how You do all things right
All things, Lord, You do right
Oh, is there anything You have not done
That was not right
Not perfect
Oh, how You rested on the seventh day
And all things were good before You
All things good before You, oh Lord
Oh, Lord, You the creator of all
You the beginning and You the end
You made all things perfect
For there is no wrong found in You
There is nothing made that was not good
Oh, how complete You are
Complete in all You do
Good in all You do
You gave beauty to all
There is no ugliness with You
There is nothing You created
That was not perfect in all its ways
Oh, how You formed us
Made all things good
And beautiful
All things, Lord
Oh, we Your creatures
You made us beautiful, Father
We must not look at the outside skin
For our beauty lies within
Our beauty will radiate out
For all things He created
Are beautiful
For He cannot create ugliness
For all things made
Are good and beautiful

Oh, how we all fall into sin
Sin, to know right from wrong
Oh, we all chose from time to time
To go wrong
Fall into sin
Oh, my Lord has made all things good
He will bring about good
From our fall
For He is a forgiving God
Forgiving to us who sin
Oh, we must recognize His goodness
His providing a way
To make all things good
As He so planned
Oh, there is no wrong found in Him
All things good through Him
All things good to those that see
All things good to those that see

May 2, 2004

Anger

What stirs the anger up in you, My child
What stirs the anger up
Is it found with the one you are angered at
Or is it found within yourself

Where does your jealousy come from, My child
Where does your jealousy come
Is it found within the one you are jealous over
Or is it found within yourself

Where does your lust come from, My child
Where does your lust come
Is it found within the one you lust after
Or is it found within yourself

Where does your temper come from, My child
Where does your temper come
Is it found within the one you anger at
Or is found within yourself

Where does your hatred come from, My child
Where does your hatred come
Is it found within the one you hate
Or is it found within yourself

Where does your love come from. My child
Where does your love come
Is it found within the one you love
Or is it found within yourself

Mar. 8, 2004

Answered Prayer

Oh, ye ask in prayer
Ye ask
Your prayers are answered
How often do you return to thank
How often
Oh, you put your answered prayers down to chance
To luck
Oh, have I not answered your prayers
Yet you continue on your merry way
Until your next request comes forth
Oh, do you even stop to think
That I have answered your prayers
I, only I have brought it about
Yet, your prayers answered and I am forgotten
Forgotten in the excitement of it all
Oh, My children
Where is your thankfulness
Where is your love
Oh, I hear your prayers
They go forth
Reach My ears
I send out My answer
Though time may pass
My word has gone forth
It will not return to Me void
Your answer is forth coming
Yet, when it reaches you
Your mind so occupied
You see your request come to pass
But your memory of even asking Me has left you
Gone from your memory of even asking
I have not been thanked
I who brought this about
I alone have brought it forth

Oh, children
You are so thankless
So thankless you are

April 16, 2004

Anything New

Oh Lord, was there anything new to You
As You stepped down from Your throne
And entered upon the earth
Was there anything new
Oh Lord, did You encounter our weaknesses
Our struggles
Oh Lord, You stepped down to save us
And, oh Lord, did You not see
And encounter the weaknesses of man
Did You not see the struggles we have, oh Lord
We can never come to Your level
No matter how hard we try
Not without help
Oh Lord, help us in our weaknesses
Help us in our struggles
For Lord, we need more faith
We need more power
We need more You
Oh Lord, help us as we travel through this life
For You know
Oh Lord, help us
For You encountered all our struggles
All our weaknesses
All our sickness
Oh Lord, help us who are weak in faith
Help us

Mar. 10, 2004

Apple

Did the apple rot on it's own
Or did something bring it about
Did the rot start on the inside
Or did it start from the outside
What entered it to create the rot
Will it consume it in it's entirety
What happens if just a piece is removed
Will it continue to rot
Or will it be made whole
What happens when it sits with the perfect ones
Does it get made whole again
Or does it spread it's rot to the others
Oh, how does one get the rot out
Can we bring it about on our own
Or do we need help from the One who created it
And what of us
Are we the same

Mar. 9, 2004

Apron Strings

Oh, how we are tied to the apron strings
Of old teachings
Oh, some teachings were good
Others not so good
Teachings that were not of Me
Teachings that you are still tied to
Teachings that hinder your very walk with Me
Oh, you learnt well you did
You student of life
Oh, you still are tied to those old teachings
Oh, afraid to really show any emotion of Me
Afraid to raise your hands in front of Me
Afraid to really sing out your praises
Afraid to dance before Me
Oh, what you learnt, you learnt well
Oh, those strings are so knotted
Holding you back from really walking close with Me
Oh, you know Me
Oh, you walk with Me
But I want you to really walk with Me
Really get to know of Me
Oh, I want you close to Me
Really close
Oh, cut those strings that hold you
Oh, you have been tied to the apron strings too long
Oh, cut them
Run with Me
Oh, run with Me

Mar. 27, 2004

As Adam

Oh, as Adam walked in the cool of the night with his Lord
What did they talk about
What was in their thoughts as they walked
One to one
Oh, what would we talk about if we walked the same walk
With the Lord
What would our conversation be
Would we tell Him of all our aches and pains
Our upset lives
Our problems
Or
Would we just walk hand in hand
And talk, friend to friend
Father with child
For He would be telling you as a father all that lay before you
He would tell you of the sights you look upon
The plants
The grasses
The birds of the air
The animals that walked on their feet
The insects
The water life
Plant life
He would tell you of what all you could do
And what you couldn't
Oh, how He would talk of all you looked upon
He would tell you all
Oh, Adam, first man
How you fell
Weakling you were
To allow someone to pull you down
Did you not know the dangers
For you were warned
Were you not

You were asked to not taste the fruit on one tree
Just one tree out of all the fruit trees
Just one tree
You walked into a trap, Adam
And fell into it
Oh, Adam, how foolish you were
And you pulled all mankind down with you
Bringing our Lord onto the scene
Causing His death
He was with the Father
Walking with Him
And He had to step down from His throne
Take His place amongst mankind
Take back what was rightfully His
A stand He made
To claim back what was His
Oh, Adam
What you brought upon yourself
And bringing mankind down with you
Oh, if you had not weakened
Had you listened to your Lord
As you walked hand in hand
Had you not weakened
And refused the woman's suggestion
Refused it
Oh, Adam—would not all your seed be in the garden now
Would they not
But what if we were you Adam, when you fell
Would we not have being tempted as well
Would we all not have fallen too
Oh, God, forgive us for judging
For we would have weakened as well

Mar. 15, 2004

Ashamed

Oh, are we ashamed of our God
Are we ashamed to say we know Him
Oh, how about our Lord and Savior
Are we ashamed of Him too
Or is it just our Lord we are ashamed of
How often have we been found to speak forth words
Of our love for Him
How often have we been found to tell that we even know Him
Oh, we go about our life
Our daily life
How many times have we mentioned Him to others during the week
Have we spoken of Him many times
Or just a few times
Or not at all
Oh, are we ashamed to acknowledge Him before others
Oh, we cannot be ashamed of Him
For He is so worthy to be spoken of
Oh, we are guilty of hiding the fact to others
Oh, what of the people we come in contact during the day
Have they not been brought by your path for a reason
Have they not crossed your path for you to speak forth a word
Oh, how you just let them go on by
Not even a word from you
Oh, where did you hear of your Lord
Many of you heard about your Lord from others
Yet, you go about your day
Hiding your light
Hiding it under a blanket of fear
Fear what others will think
Fear of telling of your love for the Lord
Oh, time is slipping away
There will not always be time
There will not always be the opportunity that is provided to you now
Oh, many of you, sit back, as long as you are saved

That is all that matters
But you are told to go out and preach the gospel
Go out and preach the gospel
To the ends of the earth
Oh, how many of you are involved with telling of the gospel
For it is life to those that hear it
Death to those that do not
Oh, you must care for those that need to hear
You must not hide under the blanket
Come out into the light
Shine your light unto others
They walk in darkness
Darkness they walk in
They are living beings like you
Do you not care for those around you
Oh, look around you
They cross your path continually
Oh, pull them from the walk of death
Pull them away
Bring them into the grace of God
Bring them in
For they are worthy to be brought in
They are worthy

May 1, 2004

Battles

Oh, what battles we must fight now
What battles
Oh, the fierceness of the enemy
Oh, how fierce they fight
Oh, what force we now must utilize
To combat and defeat
Oh, not just to push them back,
And have them to come with more ever powering force
But to eliminate the enemy completely
To eliminate them
Oh, how they attack
Using every skill
Every intelligence available
Oh, how they come in such force
Oh, we are only one to attack a powerful army
Oh, how the general yells his orders
Oh, how the individual battalions jump to his command
Oh, how they come
Marching on to battle against one
Oh, how they move
Oh, how one can combat an army
Oh, how one can fight thousands
Oh, how they are destroyed
Never to come up again for attack
Oh, they don't even run
Don't even pull back their forces
But how they fall
For one has lifted their hand
One has lifted their hand
With thousands standing in for one
Thousands standing in for one
Oh, don't fear the enemy
For their defeat is done
With one hand

With one hand
With thousands standing in for one
One with thousands
Oh, fear not the enemy for what he can do
For greater is He that fights your battle
Greater is He

April 20, 2004

Be Careful

Be careful how you walk, My child
Be careful how you walk
Watch out that pride does not trip you—
And quickly enter in
Continually guard your walk, My child
Continually guard your walk
For if pride is found in you
Your gift may be removed

Mar. 9, 2004

Behold

Behold, I am about to do a new thing
I am about to open up new doors
Within these doors are many avenues
You will be led to which door is for you
You will be led
Follow my voice
Open that door
There awaits you a new thing
Not like you have experienced before
Go boldly in
Walk boldly through
For I am about to do a new thing with you
I am about to do a new thing
Await
For I have heard your cries
I have heard, said I

Mar. 16, 2004

Be Wary

Oh, be wary of what your eyes see
Oh, be wary
For what is fact and what is fiction
Oh, can you decipher which is which
Oh, and what of the sights you see
Are they just glanced at and gone
Oh, they go into the very depths of you
Plants itself there, it does
There it grows
Just a glance is not enough
The need for more becomes a desire
Oh, you think you can cast away what you see
It doesn't have any effect on me
Oh, they move into your very heart
And grow
Oh, watch what you see
Oh, the air waves are full of visions
Full of evil
Nothing man has imagined cannot be seen
Oh, the air waves are full of man's imaginations
Man's imaginations, made into form
Oh, sent into the skies they are
Oh, be careful what you see
Oh, sift it
Sift it
Sift it
For you will become trapped by what you see
Trapped
The way out is hard
Oh, what are you looking at, My son
What does your eye see
Oh, I want you pure, My son
Pure
Stop what you are watching

Stop
Don't weaken yourself
Don't weaken

Mar. 25, 2004

Beneath the Skin

Oh, what dwells beneath the surface of our skin
Dwells within the recesses of our body
Oh, what names are they called
To those that dwells with me
That move me in all I do
Oh, what are their names
Oh, how many are there
That dwell within me
Oh, my very actions
Tell me something
Gives me an idea
Of their given names
Oh, not only I would know
But others too, would have an idea
Of their names
Of those that dwell with me
Oh, what do we see
Within our actions
Of their names
Would we be ashamed
Shocked
Deny such a name
Would live within me
Within ME
Oh, how many of us would deny such names
Deny that they just couldn't live within my walls
How many of us would be proud of many of the names
How many of us would start casting out those names
That embarrass us
That show the real me
Yes, the real me
Made us of those that dwell within my body
That have set up residence within me
Oh, observe your actions

Observe your mouth
You will know who dwells within you
And start casting out that which you wish not to live with you
Start casting them out
Call upon your Lord God
To fill you with His glory
To fill you with His love
And you will know their names as well
For, your actions will give you their names
Names that others too will know
For your mouth will tell the secrets
Your mouth will give away the real you
Oh, are you proud of what riddles out of your mouth
Are you proud

May 17, 2004

Bitterness and Hatred

Oh, don't let bitterness riddle your mind
Don't allow it to take hold and settle in, My child
For it will root itself within you
Spreading itself like a web throughout your body
It will bring about sickness to your body
Oh, bitterness is an evil thing
Bringing separation of loved ones
Bringing hate one for another
Oh, it is an evil thing
Oh, bitterness shall not enter into My Kingdom
No, it shall not
Oh, don't walk around carrying the heavy load of bitterness
Oh, it sits upon your face, allowing all to see it
Oh, bitterness is a hateful thing
Oh, get rid of your bitterness
Don't carry a grudge from year to year
Month to month
Day to day
Hour by hour
Oh, your bitterness has taken all your joy
Robbed you of your happiness
Robbed you of your friends
Oh, forget the past
It is gone
Never to repeat itself
Oh, those days are behind you
Oh, start afresh
Forgive the one you hold a grudge with
Oh, get rid of your hatred
Oh, forgive that one
As you have been forgiven
Your forgiveness was complete
Oh, can you not forgive
Oh, do it now

Do it now
Whilst there is time
For sickness and death await you
Oh, let in the light,
Get rid of that darkness
For it is an evil thing

Mar. 24, 2004

Bitterness

Oh, bitterness
Where is your root
Where did you implant yourself and take hold
Stretching your root deep in to take hold
Where only a miracle can pull you out
Oh, where did you come from
Oh, bitterness, you are of the evil one
Planting yourself into God's children
Planting yourself deep
And once planted deep
You start your growth
Invading the very soil from where you root yourself
Oh, you are watered daily
Daily you are nurtured
Daily you are fed
Oh, how you grow
You take hold like a cancer growing
Encasing yourself
Eating you cell by cell
Disfiguring as you go
Oh, how bitterness finds a home
Digs deep into the soil
Once watered and fed
It grows
Eats its way through one
Eats them up
Eventually takes over
Brings along those that ally up with it
Oh, how bitterness sidled up with anger
Sidled up with hate
Sidled up with insecurity
Oh, how they take over
Whilst we stand around and do nothing
Welcome them in

Welcome them in we do
Oh, it eats away
Eats our very loved ones
Eats our family
We are left with nothing
An empty shell
For we allowed bitterness to take hold
Allowed it to eat away at us
Oh, how we cannot forgive
We cannot let go and forgive
So, we continue and wallow in our self pity
Oh, woe is me
Oh, no one has ever walked my walk
No one has ever worn my shoes
Oh, woe is me
Woe is me
Yes, woe is you
If you remain with bitterness
Woe is you to continue with unforgiveness
Woe is you if you allow your bitterness to ride you
Take hold of you
For you will only be left with your bitterness
You will walk alone
And no one can enter My kingdom
Dragging their baggage of bitterness
Unforgiveness
Anger
Oh, let go of past hurts
Let go
For all have had their hurts
At one time or the other
Oh, you are not given more than you can bear
Oh, you must step out of them
That pull you down
Pull you away from Me
You must step out
Walk in a new light
For My arm is not too short to forgive

My heart not too small to love you
Oh, step away from that which pulls you down
Step away from that which is out to rob you
Rob you of your family
Your friends
Your life
Your eternal life
Oh, what is the cost of this bitterness
Oh, consider the cost, My child

May 15, 2004

Books

Oh, My people
Get back into My word
Get back into My word
Your time is taken up by reading well-meaning books
Books of uplifting
Books of teaching
Books of learning
Books that are written by well-meaning children of Mine
Well-meaning
I say
But you have neglected My word
You have put My word aside whilst you sat and read other books
Well-meaningful books
Yes, those books you will learn
Yes, those books tell of Me
But I want you to get back into My word
Is not all you need, found within the pages of My word
Is not all you seek found within My word
Is not all the help you need found in My word
Oh, My children
You are being led quietly away from My word
Led quietly away
You have put My word under all your other books
My word is all you need
My word is there for you to learn
To learn of Me
There is nothing too difficult in My word that Holy Spirit will not teach
For He will teach you
Lead you through what you find difficult
My word is all you need
Put away your meaningful books
Your books of learning of Me
Your books of teaching of Me
Put them away

For all is found in My book
All you need to feed your appetite
To give you life
To give you health
To give you all
For I have an abundance of knowledge for you
You only have to seek it out
Seek and ye shall find
Seek , and ye shall find
All will be found within the pages of My word
Sit and read
Sit and read

April 2, 2004

Bought

Oh, how the deceiver is out to deceive
Out to deceive
Oh, how angry he is
How he is out to destroy
To come upon My people
Oh, how angry he is
Now out to use his every deceivement, every power available
Out to attack the children of God
Oh, how he is out to attack
Oh, how he is out to make a slight change in the wording
A slight change in the truth
He is out to attack
To bring down
Gods' children
To bring them down
To weaken their faith
To make them wonder where their stand is with God
Oh, how he is out to deceive
Oh, the master of deceivement
The master of lies
The master of evil
Is bringing about all his power
All his might
Upon the children of God
Oh, how he is about to do all within his power
To deceive
To bring about slight changes from My word
Oh, how he is out to destroy
To break
To pull down
To discourage
To bring about fear
To bring about deceit of all kinds
His time is short

Short I say
Short
He knows his time nears
He knows not the time
But he knows the times
Anything
Anything
Anything—to pull My people down
Anything within his power to weaken their faith
Oh, he is there
But My children know who they believeth
They know that He is able
They know that He who shed His blood for them
Is able to bring them through
To withstand anything the evil one has brought in front of them
He is able to keep them
For all are His that have His name
All are His that bear His mark
None will be snatched from His hand
None will be taken
None
All are protected from the evil one
Oh, evil one, your ways are void
Your ways touch not My children
For all is lost
All is destroyed
All is gone
For we know Our Lord is able to keep us
Until we all come into His kingdom
For we have been bought
Bought before the foundation of the earth
Brought with His blood
Bought I say
Bought I say
Bought by the One
Bought by the Son
Bought by the Lamb
Bought I say

Bought
For we bear the mark of the Lamb
All are marked
All are numbered
We bear His name
We bear the name He has given us
All who call upon His name
Are protected
Bear His name
Are saved by His blood
All are His
All, I say
All

April 6, 2004

Breath

Oh, Lord
Did You not breathe the breath of life into Adam
And he became a living being
And is it not the breath of You, my Lord
Then carried within the seed itself

March 10, 2004

Brilliance

Oh, if I have brilliance—where did it come from
Was it of the books I read
Was it from my parents I learned
Or is it of the Lord

Oh, if I have health, my Lord
Was it of the food I ate
Or was it of my family genes
Or is it from the Lord

Oh, if I have beauty
Or a gift of such
Is it from my genes
Or is it of the Lord

Oh, if I have comfort
Is it of the work I did
Or passed down from my family
Or is it of the Lord

Oh, if I stand in a powerful place
Is it of my doings that I stand there
Or placed their by the people
Or is it of the Lord

Oh, who can say their gift was of their own
Inherited
Earned of their own
Or was it
Or is it, of the Lord

March 9, 2004

Curses

I will smite him who curses
I will smite him who curses My name
I will smite him who curses My name
For My name is Holy
Oh, what curses you bring upon yourself
Oh, ye who curse My name
What curses you bring upon yourself
Oh, ye who curse My name
What curses you bring upon yourself
My name is Holy
Holy is My name
Holy is My name

April 8, 2004

Daily

Oh, daily we are being fed by the evil one
Daily he is implanting himself
Through what we see
What we hear
What we read
Daily being fed
We are becoming immune to what we hear
What we see
What we read
Becoming immune we are
Not long ago, we would have been shocked by what
We see
What we read
What we hear
But now, we are becoming cold
Becoming blind
Blind we are
Blind to what evil is doing with us
With our children
Oh, how the evil one is implanting his evil
Implanting his evil through out the world
Oh, what violence we see
What violence we hear of
What violence we read about
It passes through our mind
Passes through
But with each passing
More and more we are not affected by what we see
Hear
Read
We are becoming more and more immune to
Sexual scenes we see
Read
Hear

Ever becoming immune to it
Ever accepting what we see
Hear
Read
As the norm
Years back, we would have not accepted it
Now, we pass it by
When are we going to step out
Step forth to the podium
Speak out for our children
For ourselves
For evil is implanting itself daily
Whilst we stand by
Weaklings we are
Weaklings we are
Evil will increase
Increase beyond our imagination
Will we still stand by after our children have been so infected
Will we still stand in the corner and allow it to pass on by
Whilst we turn our head
Oh, how evil has slid in
Slid in through the back door
And the children of God have turned their back
Turned their head and accepted it to pass on by
Read My word
What is an abomination to Me is an abomination
You are My vessels
Unless you step forth
Claim what is rightfully yours
This evil will continue to raise its ugly head
Raise its head and capture your very seed from you
Oh, when are My people going to step out
Oh, you weaklings
You are My vessels
You are My vessels
Step out
Step up
Speak out

For evil is spreading itself every second
You sit back
Every second evil is spreading, widening itself
Whilst you sit back
Sit back
I say

April 1, 2004

Daily

Oh, daily we must present ourselves unto God
Daily present ourselves
Come to Him with thanksgiving
For we must be thankful for His goodness
Oh, we must renew ourselves
Daily
Daily come before Him
Daily come to Him
Oh, we must come with a pure heart
Pure mind
Before Him
For He is so worthy of our obedience
So worthy of our purity
Oh, we must daily bathe ourselves in Him
Daily bathe ourselves in Him
For He is so worthy of our obedience
Oh, we must walk in His Word
Oh, we must come to His gates
With thanksgiving
So thankful for His goodness
His mercy
Oh, cleanse ourselves
Put on a new robe
Come into His courts with praises
Oh, come into His courts with praises
Oh, enter into the Holy of Holies
For we are cleansed
Purified
Washed with the sacrificial blood of the Lamb
Oh, our sins washed away
Oh, cleansed from all our sins
Bathed in His mercy
Wearing the robe of righteousness
Oh, we can enter into the Holy of Holies

We can come before our God
Our Saviour
Our Redeemer
Oh, how we can come before Him
Bow down before Him
For He is so worthy of our praises
For He alone is God Almighty
Cleansing us daily
Washing us moment by moment
By His mercy
His love
Oh, He is so good
Our God is so good
Oh, we must not come into His presence
With sins upon us
Oh, come to Him for forgiveness
For surely He is able to wash them away
Oh, cleanse yourselves
By the sacrificial Lamb
And come into His presence
Cleansed, purified, washed
Oh, we enter His presence cleansed
Cleansed by His mercy and goodness

May 9, 2004

Dance a Dance

Oh, ye children of the Lord
Oh, ye children
You will be dancing upon the streets
Singing upon the streets
Praising upon the streets
Playing your timbrels upon the streets
Days of joy
Days of happiness
Dancing on streets so clear
Happiness upon your faces
Joy upon your faces
No more sadness
No more tears
All are coming to worship the Lord
All are coming to see the King
Oh, coming from far away places
Coming to worship the Lord
Oh, there they come
Prepare for them
Oh, they are coming
Oh, praise the Lord
Oh sing, My children
Oh praise, My children
See the children coming
Oh, numbers one cannot count
See them coming, from faraway places
Oh, Praise His name
None missing, all are accounted for
Oh, not one lost
All are here
Oh, dance before the Lord
Sing upon the streets
Oh, praise His name
For He is worthy

Oh, praise His name
For He is worthy
Oh, praise His name
For his is worthy
Worthy is His name

March 23, 2004

Dance

Oh, dance before the King of Kings
Clap your hands with joy
Sing out aloud your praises to the One
Sound aloud you musical one
Oh, sing and shout His name
Gory, Glory, Glory is His name

March 2, 2004

Darkness

Darkness covers the earth
Cries throughout the lands
Smoke arising
Darkness
No light to shine
Hunger throughout the lands
Diseases run from shore to shore
People running to and fro
Death given over for food and water
Where are the birds of the air
Where are the four-legged animals that roam
The fish of the sea lie dormant
Woe to earth
Woe to earth
Where does one run
Where can one hide
Instant death would be welcome
Oh, where are the children
Where are the children
Cries go forth for their offspring
Oh, where to hide
Where to hide
Woe to earth
Woe to earth

March 18, 2004

GUEST QUARTERS®

S U I T E H O T E L S

For Reservations Nationwide call **1-800-424-2900**

ASHAMED P.23 (SO TRUE)
BITTERNESS P.37 (SO TRUE)

Days Ahead

Oh, that day will come
When we least expect
Oh, that day when peace is about us
Oh, that day when we dance with the bride and groom
Oh, peace abounds throughout the lands
Oh, a day when we least expect
Oh, who is prepared for such a day
Oh, who will be ready
With their robes washed white
Their lamps all trimmed
And filled with oil
Oh, how that day will appear quickly
A day when we least expect
Oh, we will all be about our life
Then, we who are the Lord's will be taken
Oh, to those left
Oh, to those left
For times and time will come upon you
Oh, children
All My children
Oh, be ready
Oh, be prepared
For such a day
Have you not been warned
Told of such a day
Oh, to you who remain
Oh, to you
Who scoffed at My word
Oh, don't let the deceiver confound you
Oh, don't let him deceive you
Recall your teaching
Your teaching of years back
Recall them
Hold fast to them

For days will come of trial for you
Days of terror
Days of weeping
Days of hunger
Days of an eye upon you always
But, oh, recall your teachings
Oh, pull them out of the drawer
Oh, listen to the words
Words of salvation
Words of salvation
For days will come upon you of many trials
Many fears
But oh, come to My word
Remain in My word
Remain in My word
For My word is true
My word is life
My word is all
Oh, remain in My word
And thou shall be saved
Oh, remain in My word
And thou shall be saved

April 25, 2004

Days of Shortage

Oh, there are days ahead
Days of shortage
Days of dwindled lighting
Days of dwindled heat
Days of dwindled food
Oh,
Days ahead
Where food is scarce
Where heat is scarce
Where light is scarce
Oh, what of those that have no light
Those that have no heat
Those that have no food
Will you share your food
Your light
Your heat
Oh, those days are nearly upon you
What will your stand be
What will your stand be
Will you keep for only yourself and your loved ones
Or will you be willing to share what you have
For to share with others is to share with Me
Oh, be watchful of how you act in those days coming
For it is easy to share with the needy when you live in a day of plenty
But what of the days of scarcity
Will you be willing to share your little light with them
Your little heat
Your little food
For those days are about to appear
Oh, take a stand now where you stand
For to share is to share with Me
For the poor have always been at the doorsteps of the plenty
Oh, have you taken a stand and shared with the needy
Or have your turned a blind eye

For your refusals may have been not noticed with man
But they have been noticed by Me
Oh, you have, because I have given it to you
I can remove it and give to the needful one
Oh, beware, lest you be the beggar at the doorstep of the wealthy
Beware, for those days are about to appear
I see my beggars
I care for them
My heart goes out to them
I ask you to care for them as I care for them
They, I worry about
Care about
I ask you to share when those days come
Share your very last morsel
Share, I ask

March 26, 2004

Death

Oh, how death sneaks itself in
Takes away those we love
Those we walked with
Laughed with
Oh, how it sneaks itself in
Removing from us those we love
How we hurt with the loss
Oh, death
Brought about by the weakness of Adam
Adam
To whom God gave authority over all
Over all he gave authority
Oh, are we any different than Adam
We too have authority
Authority over all on earth
We too walk with that authority
Authority over sickness
Authority over death itself
Oh, are we any different than Adam
Oh, we are a weak people
Giving the authority
Yet, remain still spineless
Oh, we must step out in our authority
Authority over death
Yes, over death
Oh, we too can call back
That which has been taken away
We too have that authority
Oh, evil one
Where is your bite now
Where is your bite
Oh, you weakling evil one
You are a weakling
A weakling, using your measly powers

Powers you deceived with
Oh, you are not powerful
You are not powerful
For you are a deceiver
Out to deceive anyone you can
Oh, we will not stand for your deceivement
We will step forth
And plant you under our feet
We will tramp you down
You will be trampled under our feet
Under our feet you will remain
Oh, evil one
You have done your damage
But your damage is broken
You will not remain in your strength
You will be found under our feet
For we have One who fights for us
We have One who stands for us
We have One who fights our battles
Who gives us strength
Who gives us power
Oh, we have One who is all
We have One who is all
Oh, we have Him within us
Oh, we have Him within us
To fight our battles
To give us power over evil
To give us power to put down anything that comes against us
Yes, death will not reign over us
For death is of the evil one
Oh, where is your sting now, oh evil one
Oh, we have power over you
We have all power over you
For we have One in us
Who will fight for us
Who will fight for us
Oh, we have Him within us
To walk with us

To strengthen us
To give us all power over evil
Oh, we can do all things
Through Christ who lives in us
We can do all things through Christ who lives in us

May 2, 2004

Desires

What riches do you desire
What can you purchase with it
Material things
Happiness
Sexual desires
Will these all fulfill your dreams
Can this money fill the empty gap you have
What do you seek
Is this your deepest desire
Can fame fulfil it
Can love fulfil it
Can luxury fulfil it
Oh, where would you travel to find this happiness
Where would you seek it
Will it be across the seas
Will it be within the skies
Across the lands
Oh, where will your travels take you to seek this desire
Can all your money find the end of your rainbow
Can all the money in the world—can it fulfill your desires
Can it
Oh, what can money buy
Can you obtain true love
True happiness
True life
What about eternal life
Can your money purchase that
How much will it cost you
Will you have enough
Who would you negotiate with for an entrance into this eternal life
Can you obtain it through your good works
Can money buy you a ticket
Or must you give your life over
Which way would you go to deliver "your life"

One ticket is free, paid for
The other you pay for it
Which way will you go
Think hard, My child
Your life is in the pending of your decision

March 6, 2004

Did You Not See

You were there as I lay within my mother's womb
You looked upon my feeble body
You watched as I was cast from my mother's womb
You saw my growing years
My hurts
My happiness
My loves
My deaths
You were there as I stood and laid my parents to rest
Oh, Lord, you were always there
You were there when my tears wet my pillow
You were there when I loved
Oh, Lord, you have seen my every move
You were always there
You've seen my sins
You have cast them all away
Oh, what love You are
What love You have
Oh, Lord pull back your hand from what you are about to do
Oh, give your children another chance
Father, You say You wish none to perish
Oh, Lord, lengthen our time
Show us the way
Have mercy on us
Don't destroy all that You have made
Oh, Lord, have compassion on us
How long shall I hold back
How much more shall I give you You say
Your sins are great
There is decay everywhere
How long shall I hold back
My wrath will come
No time will be allotted
For the sins of the earth reach the heavens

I will come
I will destroy that which I have made
There will be no pulling back what I have intended to do
Prepare ye people
Prepare yourselves.
Oh, Lord, do not the heavens declare Your glory
The seas even obey Your voice
Look upon this land You made
Look upon all You have made
Oh, Father, surely there is some good in your creation
Something to be found that pleases You
Oh, Father have mercy on us
We have gone astray, bring us in
We are a foolish people
Find some good in us I pray

February 23, 2004

Dirty

Oh Father, how dirty we can be
How dirty
Oh Father, is it the clothes I wear
That dirties me
Oh Father, is it the food I eat
That dirties me
Oh Father, is it what my eye is cast upon
That dirties me
Oh Father, I was dirty before you
Oh Father, what then if not what I eat, nor what I wear, nor what my eye is
cast upon that dirties one
Then, what Father
Will soap and water cleanse me
Oh Father, what
Oh—child—it is not the food you eat
Nor the clothes you wear
Nor what your eye sees
Nor the dirt of your skin
But—it is what lies within your heart
What your heart produces
And is spewed out your mouth
This is what dirties one before Me
Oh, cleanse your heart
Cleanse it
No soap and water will cleanse this
No soap and water
Oh child
Is it not the Holy Spirit that will wash over you
Cleanse you inside and out
Prepare you for Me
Oh child
Get washed
Get washed

Prepare yourself to be presented before Me
For you must be clean to be presented before Me
Clean I say

March 23, 2004

Disease

Oh, how disease will run rampant
Oh, how it moves
Moves upon the nations
Claiming lives here and there
Oh, how it moves
Powerful ones
Ones not to be combated easily
Oh, how they move
Upon one country to another
Oh, man is at a standstill
With wonder of it all
Oh, to combat it
How to attack such a thing
Oh, how steadily it maneuvers itself
Claiming life after life
Oh, new and newer they become
Until man is at a standstill
How to attack such a thing
Oh, diseases like never before
New and newer they become
Stronger and stronger they become
And man is unable to attack
Man is unable to attack
Oh, how disease has its way
Running across the lands
Bringing down man as it moves across the lands
Oh, taking the very young from under their roof
Oh, taking their very loved ones from under their shelter
Oh, where can man run to for advice
Where can they run to for help
Oh, nations come together for advice
Oh, nations come together for help
But there is no help from the nations
No nations can outwit the newer and newer strains that come

Oh, nations are unable to keep up
To the cleverness of the evil one
Oh, nations will seek help
They will seek help
But there is no help on the way
Oh, they do not seek the God Almighty
They do not seek His help
They do not seek His advice
And so they succumb to the evil one
Succumb to his outputs
Oh, how clever evil is
Branching out across the lands
Claiming lives as he moves
Oh, destruction he brings
And man still does not cry out for help
Still curses the God of the universe
Still curses the God of the beginning
And the end
Still curses Him
And still cries out for help
Help from the nations
Oh, when will man learn
When will they learn
Will they learn after they have lost their household
Will they then turn to the only God there is
Will they
No, they continue to curse His very name
They continue to curse Him for their losses
Oh, when will man learn
When will they learn

May 15, 2004

Do Not Fear

Oh, why is fear abounding
Fear of this
Fear of that
Fear of disease
Fear of the skies
Fear of the environment
Fear of terror
Fear of murder
Fear of robbery
Fear, fear, fear
Oh, fear everywhere
Daily you hear of fear
Fear in your city
Fear within your neighborhood
Oh, fear, fear, fear
Oh, slow down, My child
Slow down
Calm yourself
For I am in control of you
Do not concern yourself
For I am in control
Do not fear
For those who walk with Me there is no fear
Your fear is within yourself
For with Me, there is no fear
Oh, slow down
Calm yourself
For I am here

March 18, 2004

Do You See

Do you see the King
He is high and lifted up
Oh, do you see Him
Is He not round about us
Is He not within you
Oh, He is all about the earth
High and lifted up
Oh, can you see
Oh, open your eyes
Has He not in times past prompted you
Has He not nudged you
Has He not quickened you
In times of need did He not come and present Himself
Show His love
His mercy
Oh, you foolish generation
Open your eyes
Open your ears
Open your heart
He is all about you
He is your helper
He is your healer
He is your deliverer
Oh, take time for Him
Run into the quiet room
Run to your closet where no man may enter
Shut the door
Remain
Remain
Remain
I will come to your needs
I will present Myself
You will know of my presence

Sit back
Await
Await
Await
I will come
I hear your prayers
I see your tears
I hear your cries
They have reached beyond the heavens
They have reached My ears
Oh, I have inclined My ear to hear
I will answer
Do not hasten the answer
It will come when times are right
I know the times
I have all things planned
Oh, wait in the quiet room
Time is of no essence
Wait I say
I will answer
Your time of need will be fulfilled
For I love you
I hear you
I will answer you
Await the time, for My heavens must be readied
Principalities you know nothing about
But I hear your prayers
Feel your need
Help goes forth to you
Be patient
Be patient
Oh, how great is our King
How magnificent You are
You hear our prayers
You see our actions
You alone know our hearts
Oh, have pity of us

Have love for us
We are Your creation
Oh, love us, Your children

March 3, 2004

Evil Presents Itself

Oh, how evil presents itself
The many faces it wears
Oh, how it tricks us
Oh, how it does
The many forms it comes in
Tricking us, tripping us
Oh, how it shows itself
Oh, it is everywhere
Everywhere
Oh, it hides itself in places one would not expect to find it
Hides itself in our homes
In our schools
In our places or worship
Oh, how it hides its face
How clever it is
How clever
Oh, one must know My word
Oh, you must keep one step ahead of evil
Always, one step ahead
Without My word, you may fall
You may be tricked
You may be deceived
Oh, evil hides everywhere
Waiting for a weakened moment
Times when your mind is full of other things
Times when you are so occupied
Times when My word is placed upon the shelf in your brain
Then he will come, in his many forms
Whichever form you are weak to
Whatever form that you find hard to resist
Then it will present itself, waiting for you to fall
Oh, be planted in My word
Keep it in the front of your brain
Where you can reach for it quickly

For the evil is one step behind you, waiting
Oh, what ever your weakness is
Whatever it is, that is what evil will come in that form
Luring you, beckoning you, tempting you
Unless you are firmly planted in Me
You will fall, you will weaken
Oh, watch, My child
Watch the many forms evil will present itself
Oh, watch
Keep My word at the very tip of your tongue
Keep it at the very tip of your brain
Keep it at the very tip of your heart
That you may reach for it quickly
Before you fall

March 24, 2004

Evil

Oh, I see the evil lurking around you
Hiding within the shadows
Beckoning you
Offering all sorts of pleasantries
Oh, how fancy dressed she is
One cannot decipher whether she is good or bad
Oh, how nice her aroma is
How smart she walks
She struts in front of you
Tempting you
Does your head turn to look
Do your feet turn in her direction
Oh, how well hidden evil is
Hiding within one's flesh
So clean and pure looking
Can you tell what lies beneath that flesh
Can you know where she will lead you
What evil she can bring upon you
Are you tempted
Can you deny your temptation
How strong is your faith
Is it strong enough to resist
Can you run from just one sinful act
Can you
Oh, don't get caught within her clutches
Oh, make haste
Turn
Run
Run
Run

March 5, 2004

Faithful

Because you have been faithful with what I have given you
And walked straight before Me
To you I will bless
Bless you with an abundance
To you who have walked a special walk
To you I will bless
To you that has carried a burden for lost souls
To you I will bless
To all you that carry My word
And send it to a lost people
To you I will bless
I am about to pour out My blessings upon you who have been faithful
To you I am speaking
To you I speak
Your works have not gone unnoticed
For I have seen
I have seen your heart
I know your heart
Your works are sincere
Your works alone are not a ticket into My kingdom
But your works are for the kingdom of God
These works will I bless
These works I have seen
I will pour out My blessing upon you
Oh faithful servant

April 1, 2004

Faith

Oh, where is this faith I seek
Where can I find such faith
Faith to see my requests unfold
My hopes, my spoken desires
Oh, where must one go to locate such faith
Oh, how the servants of God of old
Oh, how they walked in faith
How they stretched their seed of faith
The seed allotted to us all
How they nurtured and watered it
To cause growth
This faith I seek
What was their secret
What was their walk
That I may obtain this faith
Oh, what must one do
Oh, the men of old
Did they not walk in this faith
Did not the maidens walk in the same faith
Oh, where can one find this faith today
Faith that can raise the dead
To heal the lame
To cure sickness
Oh, what was their answer
What was their secret
What walk did they walk
To obtain such trust
Trust in the spoken word
Trust to know without a shadow of a doubt
That their spoken word, with faith would be final
There was no second thought
They just walked in faith to know
No matter what obstacles they encountered
They just knew that their spoken word would be manifested

No matter how many years, and in fact, many not even in their lifetimes
They carried a trust in the word of God
Oh, where is that faith today
This substance we need
Oh, where must one obtain this substance
To know without a shadow of a doubt
That what we hope for
Would surely come to pass
To walk forward, no turning back
To know that it is done, before it was
Oh, I seek that faith
I seek that walk
I seek that sureness
That ultimate reality
That my spoken word, in faith too will come about
That I too, will walk the same walk
To go forward
No turning back
To know that my hopes, will come about
And my faith too
Will be blended in with the servants of old
That I will walk the same walk
The same faith
For My God is great
He is ever so faithful
To nurture this seed I have
This little seed of faith
My allotted amount
Now sprouted
Now growing
For the dew from heaven is watering this sprout
Watering it daily
Growing this plant is
Growing it is
Oh, how it has grown
My faith, now about to bear fruit
Now about to bear fruit

April 9, 2004

Falling Away

Oh, many will fall away from My word
Many will fall away
Oh, plant yourself firm in My word
Plant your feet firm
For many will cool in days ahead
Many will cool
Oh, don't be found wavering in these days
Don't be wavering
Oh, plant your feet firm
For the great deceiver has stretched forth his net
Oh, many of you will waver
Oh, don't fall away, My children
More now than ever
Plant your feet firm
For many of you will waver from your first love
Many of you will not be found in My word
Oh, don't let the busyness of the day take over you
Oh, don't let anything part you from your first love
For many of you will fall back
Many of you will fall back
Oh, plant yourself firm
Oh, don't be found sliding away
Oh, more now than ever
Get firm
Plant your feet solid
Oh, cement yourself in My word
For the deceiver is out with his net
Oh, don't allow him to pull you aside
Oh, My children
Don't fall away
For difficult it will be for you to unsnarl yourself
Difficult it will be to unsnarl yourself
Oh, stay firm
Keep in My word

Plant yourself in
Oh, ye must be well rooted
Well rooted
Or you will be pulled out
Oh, don't dry out
Keep yourself saturated
Well fed
Let the light keep you growing
Don't wither away
Oh, plant yourself firm
Don't fall away

April 18, 2004

False Teachings

Oh, My children
Don't fall into false teachings
Oh, be wary
Read My word
Study My word
For false teachings go forth in these days
Teachings that could fool one if not planted in My word
Oh, false teachings are about you
Oh, plant your feet firm
With one who teaches My word
Who testifies of Me
Who teaches My complete word
With nothing taken
With nothing added
Oh, do not be deceived, My children
For the beginner of deceivement has made his plan
A plan to snare My children
A plan to place his words out there
To snare and pull in my children
Oh, My children
Firm yourself in My word
Oh, place My words into the very depths of your being
Learn of Me
Learn of Me
Do not be deceived
Remain with one who teaches My word
For he will teach you My word
He will not be afraid to speak My word
He will not be afraid of what man can do
For he is planted in My word
He will not waiver with the winds that move
He will be planted firm
Remain with him
Learn of Me

Stay in My word
Learn of Me
Do not be fooled in these days
For it is you, My children, he seeks with his deceivement
It is you he seeks
Oh, remain with one who teaches of Me
Remain with him, My children

March 24, 2004

Famine

Oh, famine is attacking My children in distant lands
Oh, famine is attacking
Oh, this is only a beginning of their famine
Oh, whilst they die for lack of food
You, My children in a land of plenty
In a land of plenty
Walk around with your bellies full
With your bellies full
Your dogs under the table eat better than My children afar off
Oh, what of them
Oh, what of them
They die continually for lack
Lack of food
Lack of warmth
Lack of medicine
Whilst you in the land of plenty
Die from excessive amount of luxuries
Oh, where is the love, ye people of the land of plenty
Oh, where is your love
Oh, ye treat your animals better than you treat your fellow man
Oh, ye, who dwell in the land of plenty
You have been blessed
Been blessed
You have been blessed because you blessed Me
You placed Me upon you nations
But slowly you are removing Me
Removing Me from within your nation
Removing Me from your governing rule
Taking Me off the mantle
You are
Oh, ye, who have plenty
Beware, for famine too will stretch forth unto this land of plenty
Famine and disease sit upon your doorstep
It will flood your land

Oh, whilst you stood back
Watched My children die in distant lands
Stood by, turned your head
While they cried themselves to sleep
Your dogs under the table reveled in your abundance
Oh, while you had plenty
You walked with pride
Pulled Me down
Removed Me slowly
Allowed the evil one to enter fully into your nation
Oh, disease and famine you too will taste
You will feed of your dogs
Them you will feed of
Oh, ye selfish nation
I will pull you
I will pull you
Oh, to consider your dogs
Better than My children afar off
Oh, whilst they die for lack
You walk with gluttonous feet
You gluttonous people
Your brothers and sister afar off
Your mother and fathers
You treat your animals better than them
Oh, ye selfish, gluttonous nation
Oh, come back to your roots
Come back to your beginning of a nation
Where love abounded
Where I was placed within the founding of your nation
I was in your rulers quarters
Oh, come back, My children
For famine and disease will spread over your land
For you have opened its door
To come in
To come in

April 13, 2004

Father

Father, what do You say to me in the quietness of the night
What do You say
What words have You implanted within my mind
What words within my heart
Oh Father, your mystery amazes me—implanting thoughts within my mind
Oh Father, what words you have given me
Surely not words of mine
Oh Father
How mighty You are
How do You sneak the words into my mind
How do You do it
For my mind is afar off
Yet I hear Your words
For You have given me words that totally mystify me
You've given me thoughts that I myself have to check my own heart
Of where I stand with You
And Father, what of those that read the words
Do they not check themselves

March 8, 2004

Fear Grips You

Oh, how fear grips My people
Fear of this
Fear of that
Oh, have I not said to fear not
Yet, you fear the pestilence that comes your way
Fear of terror
Fear of the air you breathe
Fear of the catch of the seas
You fear the waters
You fear the food you eat
You fear constantly
Oh, you run like a chicken with no head
Run here
Run there
Seek this
Seek that
Oh, is there no end to your nonsense
Is there any end to your fears
Have you not been told to not fear
The terror of the night
Nor the arrow that flies in the day
Or the pestilence that walks in the darkness
Or of destruction at noon day
Oh, is it not so that 10,000 may fall at your right side
And a thousand to your left
And it not come near you
Yet, you fear, fear, fear
You fear, fear itself
Fear the Lord God only
Fear the Lord God only
Is it not said to fear the Lord is to love Him
Your Lord God
Oh, fear not
What tomorrow brings

Have I promised you a tomorrow
Oh, fear not what you eat
Does not your blessings cover that
Oh, fear not the skies
The earth
For they are mine, says the Lord
Oh, does not My hand guide you continually
Does not My eye
Watch you continually
Oh, are you not the apple of My eye
Oh, fear not what comes upon the earth
Fear not what comes upon man
Oh, fear not the movement of the skies
The shaking of the earth
For I am with you
I will shelter you
I will hold you in the palm of My hand
Oh, My little child
Come to Me
Come to Me
Fear not what comes around you
Trust in Me
Trust wholly in Me
Put all your trust in Me
And I will guide you through
I will be your strength
Did I not guide the children of Israel
With a fire by night
And a cloud by day
Did I not give them food
Give them water
Clothe them
Keep them from being barren
Keep them from sickness
Did I not do for My children
Oh, walk in pure trust
Walk in pure trust
Trust wholly upon Me

And I will be with you
I will never leave you
Never

May 5, 2004

Fear the Lord

Oh, we must fear the Lord
For He is great
Oh, reverence Him
For He is great
Great is our Lord
Oh, how gentle He is
Oh, how loving
Oh, fear the Lord
For He is great
Great is the Lord
Oh, fear Him who made all
Who created all things
Oh, the beginning and the end
Yes, the Alpha and the Omega
Oh, reverence Him who reigns
Reverence Him who reigns
Oh, bow down before your Lord
For He is great
Oh, great is He that is in me
Than he that is in the world
Oh, fear Him
Who created all things
Oh, what joy He is
What love He is
For He is good
He is good
Oh, bow down before your King
Oh, give thanks for His many blessings
Oh, bless Him
For He is worthy
Oh, bless Him for He is worthy
Oh, give praises to the King
Give praises to the King
For he alone is worthy

Worthy of our praises
Worthy is He

April 18, 2004

Filled

Oh Lord, how You are everywhere
You are everywhere
Filling the earth
The skies
The vastness around the heavens
You fill it all, oh Lord
You are within every cell
Every molecule
Oh Lord, how You are within all
Your very Being moves within Your creations
Your very Being fills the very present
The past
The future
Oh Lord, how You are about everywhere
You fill every breath I take
You are ever filling me with Your presence
How You fill and refill
Oh Father, You are my very breath I breathe
You are my every cell
You are about me
Before me
Around me
Behind me
Oh, how You are everywhere
My very breath, is You, oh Lord
Creator of all
You are the past, the present and You are the future
Oh Father, You were at the beginning
You are the beginning
And You are waiting at the end
You are the end
Waiting to gather us in
You know our names
You know each one of us

Your name is written within us
Your children, how You know us
Each of us different
Each of us unique
Made by You
Created for You
To give You glory
To give You praise
And how our praises now will seem like child's play
To our praises to You in glory
How our praise will flow, Father
Flow to our everlasting Father
You, oh Lord, worthy of our praises
Worthy of our praises
You are

May 1, 2004

Flags

Oh, I see the flags of the nations flutter in the breeze
Flutter, flutter, flutter
Flags of this nation
Flags of that
Big nations
Small nations
All colors they are
Solid colors
Stripes
Stars
Moons
Oh, what principalities govern these flags of the nations
What hidden rulers sit perched upon the flagpole
Flutter, flutter, flutter
Who really rules the nations
Who
For it is not the leader that rules
But, he who rules him
Oh, are there not ones who sit upon the skies of the nations
Issuing orders to the leaders
Go to battle
Refrain from battle
Do this
Do that
Flutter, flutter, flutter
Oh, how bound the nations are
How bound and wrapped around they are, like a web
Wrapped up tight they are
Wrapped so tight some are
Oh, woe to those under those flags
Woe to them
Flutter, flutter, flutter
Oh, what evil sits upon those flag posts
Oh, what evil

What powers rule that nation
Stirring the pots
Casting down all evil upon those nations
Blinding the very eyes of those under its flag
Causing hunger
Causing war
Causing disease
Causing death
Causing the worship of other gods
Oh, what evil binds that nation
What rulers sit squatted above the flag post
Flutter, flutter, flutter
Oh, how the evil one wraps whole nations up
Wraps them tightly
Places them in a binder
Pulls them here
Pulls them there
Pulls them under
Oh, woe to that nation under that flag
Flutter, flutter, flutter

March 20, 2004

For Those That Have an Ear

Oh, for those that have an ear
Hear what the Lord is saying
Make haste, My people
Make haste
So many, so lost
My word must reach to the four corners of the earth
Every ear must hear
Every eye must see
My word goes forth
Oh, latch onto it
Digest it
Oh, see what My word can do
My word is powerful
My word is true
My word is life—to those that eat it
Oh, taste and see
Try Me
Oh, there will be a time when My word will not be taken
A time in the future when My word will bring about death
Oh, prepare for those days
Strengthen yourselves for those days
Days of testing for you
Days of imprisonments
Days of a casting out from your loved ones
And yes, days of death—for My word
Oh, make haste
Go forth
Speak My word to every ear that I have led you to
Their choice is theirs
But you must go—let the children know
My will is that none—none be left behind
None to be lost
I have prepared a home for them
Let the children know

Speak softly to them
Speak carefully to them
Speak My word
I will meet with them—soften their hearts
Weaken their cold hearts
But go, My people—tell of the good news
Tell them of My love
My words are life giving, healing words
Oh, taste and see
Taste and see

March 4, 2004

Forgiveness

Oh, to you who weakens
And falls into sin
What was in your heart
Came forth
You fell
Walked right into sin
Oh, to ye I speak
To ye
Oh, did not My servants from days past
Who walked closer that a brother with Me
Did they not fall
Oh yes, My child
They all fell
All of them
But
They saw their error
They saw their sin
They asked Me for forgiveness
Oh, is My hand too short to forgive them
Oh, I will forgive them who ask
I will
I will wash away their sins
I will cast it out into the deepest of the deep
Oh, it will be removed from My slate
Oh, My child
Though you fall
Though you sin
I will forgive
I will forgive
I will forgive

March 25, 2004

Fruit

Oh, what fruit have you produced
What fruit have you produced
Oh, ye dead tree
Ye stand there fruitless
Oh, ye dead tree
Where is your fruit
Where is your fruit
Oh, ye stand there, idle, useless
For what good are you
Taking up space
Taking up space
Where is your life
Oh, is the sap completely drained from you
Or is there still a trickle remaining
Awaiting
Watering
Nurturing
Or are you allowing that trickle of sap to evaporate
Evaporate from whence it came
Oh, dead tree
I speak to you
I speak to you
Water the dead tree
Nurture it
Allow the sap to accumulate
And run freely within itself
Oh, so that life will once again flow through the branches
Bringing forth shoots
New life
Leaves
And fruit in due season
Oh, ye dead tree, I speak to you
Unless you are revived
Revived I say

You will be cut back
Cut back
Oh, cut back
To start again
Whilst there is still life remaining
Oh, ye will be cut back
Ye fruitless tree
Cut back
Cut back
To restart again
Cut back
I say

April 22, 2004

Gates of Decision

You stand at the gates of decision
Why do you wait
Decide now—which way you go
I will not snare you
I will not pull you
You are the one to decide
You are the one to lean this way or that
I await your decision
I await you beckoning
I hold back
I want your choice, you decide
I have my mind made up
I know your choice
I want to hold back
You whimper and groan—
The time is not yet
My life is before me
The years spread out
I want to see your choice
You come to me—
Or you come to the world
What do you want
My bright lights draw you
My light pulls you
You are pulled in two directions
One will give you power—money—recognition
I will give you only love
You decide
Only you can decide
My gifts are unseen
Their gifts are seen
You can jump into the stream
The stream can pull you
My love is pure

My love is true
I await your call
I await your decision
I await

January 26, 2004

Gifts in Store

Oh, what gifts I have in store for you
What gifts I have
Oh, eye hath not seen
Nor ear heard
What I have prepared for you
Oh, I have a vast wealth
I have so much to give to you
Oh, your mind cannot comprehend
What I have in store for you
Oh, your house is being prepared
Your house is being built
Oh, I have so much for you
So much
Oh, when your house is complete
I will call you home
I will call you home
Oh, My child
I have so much for you
My faithful servant
So much
Oh, eye hath not seen
Nor ear heard
What I have prepared for you
Oh, My child
My child
Soon you will come home
Soon you will come home
Oh, what joy
What happiness
When My children come home
Oh, their work is done
Their time on earth complete
Oh, happy is that day
Oh, happy is that man
Who prepares himself for that day

Oh, happy is that man
Who has his heart set on heavenly things
Oh, happy is that man
Who knows where his treasures lie
Oh, happy is that man
Who has set his eye upon the rock
Oh, happy is that man
Who clings to that rock
Clings
Where the oceans roar
The wind blows
When the waves cast upon him
He is still there clinging to the rock
Oh, happy is this man
Who has had his trials
Has walked a walk of faith
Who has been tested
Tried
And found yet still clinging
Onto the rock
Oh, happy is that man
For he has found favor with God
Has found favor
Has had his eye upon the goal
His eye upon the finish line
Oh, a race well run
My faithful servant
A race well run
To you, My child
To you
I have so much for you
Your home nearly built
Your time short
Oh, happy is this man
Who comes home
Oh, happy is this man
For he has found favor with God

April 7, 2004

Gifts

Oh, what have you done with the gifts you were born with
The gift of speech
The gift of mastering a musical instrument
The gift of singing
The gift of knowledge
The gift of insight
What has been reproduced from your gift
Did you hide your gift for no one to know
Did you use your gift for your gain
Did you share this gift
What use did you put this gift you were given
The gift I gave you
I gave you the gift to bring the children to me
You were their teacher
Their mentor
But did you reach any ear
Did anyone hear where your gift came from
Oh, you greedy one
Beware—I can remove that gift
Give it to another
I will bring someone from the streets
A beggar
A prostitute
A wayward child
But I will remove your mantle and give to another you selfish one

March 6, 2004

Given Up

Oh, what have we given up for the Lord
Oh, what have we left behind
Have we really given over everything
And followed after Him
Did we give up the net full of fish and follow after Him
Did we give up burying our parent
To follow after Him
Oh, what have we given up
Left behind
For Him
What I ask
Oh, we ask for a closer walk with Him
But really we want the closer walk
And still retain our status in society
Our place in the family
Our wealth
Our friends
Oh, have we really given over all we love
For Him
Oh, what we have here is only temporal
Only temporal
Oh, we shall seek after that eternal
Oh, does wealth, family, status
Does it really matter
When placed side by side with the Lord
Oh, are we willing to give up our catch, follow after Him
Live the life of a poor man
Owning nothing
Depending on the goodness of others to feed and shelter us
Are we willing to do that
Oh, we seek the close walk with the Lord
We seek Him
A close walk
But we drag along our wealth, our family, our friends with us

Oh, we can't walk close with the Lord
And drag all our baggage with us
Oh, ye must travel light
Oh, ye with family and loved ones
Deal with your matters first
Then seek after the closer walk
Oh, ye that desire the close walk
Oh, cast aside the matters of this world
Cast them aside
For greater matters
You will be about greater matters
Your rewards greater than earthly wealth
Oh, greater it will be
Oh, ye who desire the closer walk
Deal with your earthly matters first
Then follow after me without guilt
With out resentment
Without regret
And then follow after me freely
With an open heart
An open mind
Oh, ye who seek this
Ye must give up all and follow after Me

April 22, 2004

Guilt

Oh, how we immediately feel guilt when we have wronged
When we have walked down the wrong path
Oh, how we feel that tingle, that nudge
Letting us know that we have strayed
Oh Father, how You let us know
Oh, even a little child knows when they have wronged
Oh Father, You have planted in us a homing device
That tingles as soon as we have wronged
Oh Father, how we feel it
Oh Father, how can we see others who wrong
When we have such huge ones ourselves
Help us to mend our ways
Help us to become clean
Daily washing
Daily cleansing ourselves
Daily putting clean clothes on
Oh Father, we are so weak
Help us to love one another as ourselves
Help us to help the needy
Help us to forgive
Oh Father
We have such a long way to go
Oh Father
Start with us now
Start to mold us to Your liking
Mold us to Your Son, Lord Jesus
Mold us to Him, oh Lord
Oh, we have so far to go
Start with us now
Oh, sharpen up the homing device You have planted in us
Let us see our faults
Our wrongs
Start with us now
So that when He appears

We will be like Him
We will be like Him
I ask

March 27, 2004

Hands Raised

Oh, I see the hands raised
I hear the praises sung
I see the dancing
Of those that have a pure heart
A tender heart
Oh, how I take pleasure in your praises
Your dances
Your songs
But
I see also the hands raised
The praises sung
The dancing
Of the cold in heart
A heart not of mine
Your praises
Your songs
Your dances, are blown away in the wind
Serve no purpose
Except to lift you up within yourself
Or, for others to see your purity
Oh, ye who I speak to
Have an ear
Have an ear
Come out from yourself
Come out
For your reward is only of man
There will be nothing for you
You who feign your love
Oh, ye who I speak to
Come over to the waters
Step in
Get wet
Oh, come out from yourself
Get into the waters

Let the waters flow over you
Oh, ye who have a fickle heart
Oh, your dances
Your praises
Your songs
Leave the very air that surrounds you and disappears
Going no further than the eyes around you
Oh, ye—step out
Step into the waters
Remove your cold heart
Come to Me in truth
Come to Me in purity
Then
I will see your dancing
Then
I will hear your praises
Then
I will hear your songs
I will see
I will hear
Oh, the angels will clap their hands with you
Oh, the angels will sing with you
Oh, the angels will dance with you
Oh, does not all of Heaven rejoice when one comes in
Oh how it does
Oh, what joy is displayed when one comes in
Oh, what joy
Oh, ye with the cold heart
Oh, it is you I speak to
It is you

March 24, 2004

Hate

Oh, what do you look like, you spirit of hate
What do you look like as you expose yourself
Do you appear with beauty
Deceiving the near impossible
Oh, what do you look like, you spirit of lust
What do you look like as you expose yourself
Do you too come with beauty
Deceiving the near impossible
What do you look like, you spirit of hate
Oh, what do you look like as you expose your face
Does your face show love, or does it show hate
Or what does it show
And what do you look like, you spirit of thievery
Oh, what do you look like as your thievery is exposed
Do you show a face of stability
A face of sureness
Or what does it show
Oh, what do you look like, you spirit of love
Oh, what do you look like as your love is exposed
Do you show a face of love
A face of truth
Or what does it show
Oh, who can tell, truth from fiction
Who can tell
Who can decipher what is truth or lie
Who can tell
Is there someone that I can go to help me in my conquest
That can tell me which is right
Who to believe
Which way to go
Which way to turn
Is there anyone that can lead me to the truth
Anyone
Find who is the truth, and find who is the lie

There your answer lies
Find for yourself
Which is the liar and which is the truth
Which is the Father of Truth
And with is the Father of Lying
Trace back the records
Where is the beginning of it all
There is a start somewhere
Search back you clever one
Search back

March 10, 2004

Healing

We ask in prayer for healing
Our words have spoken of the healing we require
Our words go forth
What forces do our words encounter as they travel the airwaves
What forces
Thickness within the airwaves
So thick, our words can barely slide through
Oh, what force they have to push to continue on their way
They reach their destination
An answer is sent forth
A struggle encountered again
Time is consumed
The thickness so great
Your answer arrives
What is your stand when the answer comes
Has the request been forgotten
Have you given up awaiting your answer
Or were you sincere in your request
You do not know of the spiritual forces around the earth
Can you not await your answer
Await
Await
For your answer has been sent forth
It will arrive
It will arrive
I have said

March 10, 2004

Hear My Words

Oh, hear My words
Oh, hear My words
My words go out
Reach those with an open heart
Reach down into their very being
Oh, My words touch
Oh, they cut
Oh, they search out the very pure ones
Oh, people
Listen, listen, listen
For I am doing a great work in your day
Oh, listen
Wonders will be shown
Both in the skies and on the lands
Wonders done here and there
Wonders done on the streets
Wonders done within the home
Wonders done within the waters
Wonders done within the skies
Oh, see what I am about to do
Oh, listen, listen, listen
For there will be wonders that your eye has not seen
Oh, listen, My children

March 19, 2004

Heaven and Hell

Oh, where is Heaven
And where is Hell
Is Heaven behind a veil
Is Hell behind one as well

March 9, 2004

Highway

Oh, what highway do you tread
What highway is it
Are you on a goat's path
Or are you on a gravel roadway
Do you walk upon the smooth pavement
And what of the stops you take
What stops do you linger at
What entertains you at these stops
And what of the forks in the road
Which one to take
What of the wrong turns
The wrong people we meet along the way
What influences do they play in our walk
The helping hands
The greedy hands
The people that show their love
By bringing you into their homes for eats
And what of the ones that take you into the house of sin
Oh, what walks do we walk
What have we learnt along the way
What of the lessons, both good and bad
What have we gained in knowledge
Oh, when we take a detour
How long before we find our way onto the correct road again
And the wrong fork we took way back there
Where did it lead us
Oh, to walk all the way back
Back to where we were
To start again
And what of the gravel road
Bumps all along the way
We look over to see the smooth highway
Yet, we still walk the bumpy road
Our every step a bump

Not good enough we feel to travel the smooth route
Not able to see the often turnoffs to the highway
Too blinded by our thoughts
So we continue on, along the bumps
And then what of the goat's path
A narrow walk, sharing it with the goats
Looking over at the gravel wide road
Not even seeing the highway over yonder
Oh, how foolish man is
How foolish
A smooth highway, a gravel road, a path
We walk along the path
Meeting with the goats
Or we are on the gravel road
Meeting with the wicked
When all along we could be traveling along the smooth road
Meeting the good and the bad
But learning from both
The forks in the highway, clearly marked
The detours with signs to lead you back onto the highway
Oh, how foolish man is
To be able to walk the smooth
And still is treading the gravel and the path
Oh, how foolish man is

April 12, 2004

Hoarding

Oh ye, who hoards
For what use is it
Will you find use for all you hoard
What of those that could have bought what you have stacked away
And what will you do with all you have
When you are but dust beneath the earth
What then
Will your offspring excel in your holdings
Can you take it all with you what you have tucked away
Or will it go back to where it started
Before you purchased it
A circle about to begin

March 10, 2004

Holy

Holy is the Lord
Holy is the Lord
Oh, search the earth
Oh, search the skies
Is there anyone Holy
Oh, is there
For He is Holy
For He is Holy
Holy is He
Oh, mighty God
How powerful You are
How mighty You are
Oh, Lord
You alone art Holy
You alone art Holy
Holy is Your name
Holy is Your name
For heaven and earth may pass away
But You oh Lord, You oh Lord
Holy are You
Holy are You
Oh, worship the King
Oh, worship the Lord
Oh, bow down before Him
Oh, bend Your knee before Him
For he is Holy
For He is Holy
Holy is His name
Holy is His name

April 28, 2004

How Merciful

Oh, how merciful You are
How wonderful Your ways
Your love beyond measure
How can I match Your love
How can I compare
How can I measure up to You
Your love so great
Is there anyone who can measure up—no, not one
I see your love
You dance upon the clouds
Your love I feel in the wind
What a wretch am I
What do You see in me
I am but nothing
The lowest of all
What do you find in me
Oh Lord, my God
What do You see
I am so unworthy of you
How can I match your love
How can I measure up to You
Oh, be patient with us
Be slow to anger
Our ways are not Your ways
We are a sinful people
Oh, help us for we are weak
Strengthen us for we are so vulnerable
Oh, help us I ask
Don't allow Your wrath to come upon us
Oh, how weak are our ways
How open we are to sin
Oh, help us to become strong
Strengthen us like You
Who can match up to You

Who is worthy
Oh Lord—help us weaklings
You've provided us a way
But Lord, help us to find that way
Show us the path
We know not which way to go
Guide us where we should go
Don't allow us to be steered in the wrong direction
Oh, none would wish for Your wrath
No not one
Oh, have mercy upon us
Guide us to where we should go
Pull us the way You want
Your desires are our desires
Your love, our love
Oh Lord, how weak we are
Oh, have passion
Have mercy
Have love
For we are a weak people
Lord, You see me as I lie upon my bed
You see me upon my feet
Lord, Your eyes see my every move
Your vastness is everywhere
We are as mere ants beneath Your greatness
Oh, what can compare
Do Your eyes see all our doings
Do You see any good in us
Oh Father, You gave this earth to my Lord for a footstool
Oh, how small we are
We are as mere dust beneath His feet
Oh Father, don't tread upon us
Don't blow us away when You speak
See some good in us
Search us, surely there is something in us You desire
Father, don't destroy that which You made
How intricate my parts are
How clever You are

You made my inner parts
Surely you see some good
Oh Lord, Your majesty fills the heavens
We ask to see You face to face
Who can look upon You
Who can stand after standing before You
Oh, what a foolish people we are
Love us with all our faults
Forgive us, for we sin before You, oh God
Forgive us

February 24, 2004

Humble

Oh, how we must walk ever so humble before our Lord
Oh, don't let pride step in
And remove our lamp from us
Oh, we must walk humble
Oh, when pride steps in
Your flame will start to flicker
Flickering until you fill your lamp with fuel
Oh, once your light goes out
Oh, your lamp is given to another
Your lamp
Given to you
To keep filled with fuel
Unless you tend to your lamp continually
You lamp will be removed
And handed to one
That will care and tend to it
Oh, how pride steps in
Unnoticed to you
Yet it steps in
Innocent at first
Oh, as soon as you see your lamp flicker
You know your fuel is low
You know pride is slowly snuffing out your flame
Oh, get rid of pride
For it will quench your flame
Leaving you with no light to shine
Your lamp useless
Fuel hard to come by
When pride has its way
Oh, your lamp then given to one
Who walks humble before their Lord
Oh, your lamp will be removed
Given to another
Oh, watch out for pride

Your flame leaving you in the dusk
Not ready for the approaching night
Oh, keep it ever so filled
Continually check it
Keep it filled
Keep your wick trimmed
Ready for use
Keep the glass clean
For your light to shine
Oh, watch out
For pride awaits at your door
It awaits
Always at your door

May 2, 2004

Hurts

Oh, hurts lie within My people
Deep hurts
Wounds not healed
Scars not yet healed
Oh, hurts, cuts, bruises—lie within My children
Oh, My children
How I will mend
How I will heal
For I have seen your hurts
I have cried with you
I have walked with you
Oh, My children
Lay aside your hurts
Oh, children, you are so bound
Your hurts
Oh, cut that which binds you
Cut it
Step forth into a new day
For all have hurt
All have walked a road of hurt
All not your walk
But a different walk
Their own walk
But all have a tale to tell
All can speak forth hurts
From their past
But My children
A new day has dawned
A new day
A day that I will walk you through
A day that I will show you My love
My mercy
I was always there
But your eyes so blinded by your hurts

Your failures
You did not see Me
But I will walk you through your past
I am here to heal your hurts
Heal your bitterness
For nothing is impossible for Me
Nothing
I offer you My hand
Reach out, take it
Walk away from past mistakes
Past hurts
Past bitterness
Walk away
For the steps of the past years will be washed away
Fresh new sand lies ahead
Oh, fresh new sand
Where we will walk hand in hand
Side by side
Cast your past into the ocean
Cast it away
It is gone
Can you go back to your mother's womb and start afresh
Oh, forget the past
Forget it
You have learnt well from your hurts
Your cuts
Your bruises
You have learnt well
For a new thing has come about through it
Bringing about a new you
A person I can use
A person to relate to those now walking in your steps
Oh, you can relate
Oh, reach out to the hurt
To the cut and bruised
For you, My child, can touch
Can teach
Oh, now you have eyes to see

Ears to hear
You can now touch these lost people
A people that walks your past walk
A people lost as you were lost
Oh, reach out to them now, My child
Reach out

April 5, 2004

Husbands and Wives

Oh, women, what have you done to your men
What you have done
You have brought down their manhood
Brought them down to be like women
You have made them as women, you did
You demand your rights
You demand your ways
And now you have brought your men down
To be as women
No longer are they the man of the household
No longer do they reign in the house
No longer are they the respected one
No longer do their sons and daughters look up to them
No longer do they respect their father
No longer does his word reign in the household
Oh, women, you have done this
You have brought this about
You have done this to your men
Oh, where is your honor of your man
Where is the respect
Is not the man the first made of God Himself
Did not God Himself take a rib and make woman
To be a mate unto her man
Oh, where has the respect gone

Oh, men
Where has your love thy woman gone
Where is the respect gone
Where is the ruler of the household gone
Oh, you, the head of the house
You sit back, let thy woman take over
You have sat back, rested upon your backside
Whilst your women take over
You have allowed her to take the reigns

You have sat back whilst she did all the work
Did all the ruling within the household
Oh, men, you have brought all this upon yourself
You sat back
Were you not the ruler of your household
Where you not the one to take charge
Yet, you sit back, lazy one you are
You sit back, whilst your woman take over
If she did not take over
Where would you be
Oh, you, who I made head of the household
Gave over your rights
You gave over your rights, you did
Oh, where do you stand now
You man of the household
No better than a servant now
No better than a servant
You allowed your woman to take charge
Only because you did not
She took charge
And she did well
But, you were made head of the household
Oh, where is your shame, you men of the house
Oh, where is your shame
You act as women now
You walk as women now
You feel like women now
Oh, where is your manhood
Oh, where did it go
You slid back, took the woman's role you did
Oh, men, take hold
Take hold of your rights
Oh, unless you take hold
You will have lost all your right to rule over your household
Oh, you are giving that over
To the woman
Oh, she is capable of ruling
But you were given the right to stand as head

But you slowly gave over that right
You took the back seat
And now you sit riding on the backside of a woman
You ride on the backside of a woman
You, who I made head
Now ride the backside of a woman
Oh, men
Where is your stand
Oh, woman
Take your rightful place
For I created man to be head of the household
I created woman to be a helpmate unto him
Oh, take your rightful places My children
Take your rightful places
Oh, you wonder why your children have no respect
Oh, you wonder why there is no respect
Oh, look upon yourselves
Look upon yourselves
Look at your lives, look at the lives of others
Oh, men, you have taken the back seat
Whilst your woman ruled
She worked hard
She did all
She did the household chores
She brought in the earnings
She raised the children
Whilst you sat back
Sat in the back seat
And allowed it
You men, I have a something to say
You, I made head
And you took the back seat
I hold you men responsible for where you are now
I hold you responsible for where you stand now
You stand where you are
Because you allowed the woman
Your helpmate
To take over

To raise the children
To do all
While you sat back
Rode on her coat tails
You did

May 5, 2004

Hush

Hush—sounds afar off
Hush, all the heavens
Hush, all ye on earth
Hush, beneath the earth
Hush—sounds afar off
Oh, woe to the inhabitants of the earth
Oh woe
The key is about to unlock the chains
Oh, woe to earth

March 13, 2004

I Am

I am the One that gave you life
I am the One to take that life
I created you from the dust of the earth
I will be the One that will return you to that earth
I molded you and formed you
I rolled you over and breathed life into you
I gave you life—My life
I do as I please—your life is mine
I created you before the foundations of the earth
I knew you before your mother cast you from herself
I saw your coming and I saw your going
I knew when you would stumble
I lifted you up—I placed your feet upon the ground
I gave you solid ground to stand
I saw when you sank into the mud
I pulled you up
I cleansed you
I determine what I want for you
There is nothing hidden from Me
I see your every move, every thought
I go before you
I give you balance
I give and I take
It is I that only sees the plans I have for you
Only I

January 4, 2004

I Am But Nothing

I am but nothing before you, oh Lord
I am but mere dust
Dust to be blown away
A return to the earth
You see my works
You search my works
Are You in amongst my works
Are You there with in my life
I am but a mere mortal
You have your way, oh Lord
You mold me as you want
You take me and use me
For You are my Lord
The Mighty One of Israel

February 13, 2004

I Am Coming

I am coming
I will rip through the lands
I will pull up that which I want
I will tear down that which I see
Look what is there—
Those who stand so mighty
Those who say there is no God!
Watch what I will do
There will be none left
There will be none to say there is no God
I am the Almighty
I am the Powerful
I will cut through
I will slash dead branches—
The young cry out
The old fall
None will stand up to me—
I come before you
Prepare yourselves
Prepare yourselves
There is none like me
I will move the lands
I will push the mountains
The seas cry out
The dead cry out
There will be none that can stand
I will not be gentle
I will yank you from your feet
Move over you lands
For I am coming

January 25, 2004

I Ask to See Within the Spiritual Realms

Oh, how I ask to see within the spiritual realm of God
Oh, how I long, just a peek
Just to see as others have in days past
Just to take a quick glimpse inside, what goes on
And then, I was given a quick peek, a flash only
Looking in at the amount of movement
Like an escalator going up into the clouds with angels going up
Another escalator coming down from the clouds, with angels coming down
Oh, the amount of them, some just appearing to have conversation with others, movement here, movement there.
The amount, was quite exceptional
Then
I asked for a sneak peek inside the spiritual realm of the evil one
Just a quick look to give me some idea with what we are dealing with here on earth
I was taken, for days, upon days away
Traveling miles upon miles
I was wondering where were we going to travel such an extent
I had always thought that the spiritual realm of the evil one was just above us in the skies
I was taken and taken
Traveled and traveled
Finally I was taken, to my country
My city
My street
We stopped at my address.
I looked at the number, and was shocked
I asked what did this mean
It means, for me to see within the spiritual realms
I must open my eyes
For the spiritual realms are all about us

All over
Yes, my country
My city
My street
And yes, my home.
When I see the spiritual realms within my own address, then I can see beyond
and see where else it lies
For the spiritual realms, both
Are everywhere
Everyone.
Not just up with in the skies as we would imagine
But down here, on the lands as well
And yes, at our address too
My home and yours

March 21, 2004

I Await

You stand and whimper—and cry out
Lord—Lord
I do not hear you
You cry out—cry out
Where does the cry come from
I do not hear
My ears awaited the cries
I awaited to hear
No sound came
I closed the gates
No sound came through
Your cries reach the gates
No one is there to receive
No one to care
Too late
Too late
I awaited your call
I awaited your call
No call came
The gates are closed
Your call did not reach me
Time ran out
Your life is gone
Enjoy what you have
For your life is as that
I would give you life—My life
But you chose to go your way
Go—Go—Go—
For I will not pick up the pieces
I will not be there at the end of your life—Go—
I give you My life—My dreams, My love
But go your way, go your way
I cannot and will not deter you
Your mind is made up

Your mind is your own
I gave you life—but you give the avenues to take
You give the decisions you make
You do it all
I await your decision
I await your decision
You only have to call
I will be there
I always stand in the background—
Awaiting your call
If you don't call
I can do nothing—nothing
You are your own destiny
I welcome you—
But you have to come to me
I love you—
But you must love me

January 27, 2004

I Search

Oh, I search the earth
Searching out for one to carry My message
I have searched the earth
Searched a people
Searched for a vessel to carry My word
I search out one with an open heart
A genuine heart
One willing to carry My word
My word will I give to that vessel
My word
It is not the vessel you look at
The vessel could be broken
Cracked
Worn
Parts missing
Paint worn off
Or
Perfect in its stature
Beautiful to look upon
But the outside is not what I look at
It is the inside I see
I will see beyond the flesh
And look upon the heart
For I see the open, willing heart
The humble
The weak
Them, will I choose to use as My vessel
To carry My word
Them, will I choose

March 12, 2004

I See

I see your works
I know your ways
You cannot hide from me
You are high and mighty
I will bring you down
I will break you
You will humble yourself before Me
I move across the lands
I will bring down who I want brought down
I will raise up who I want raised up
The stench of earth reaches the Heavens
I cannot hold My anger for long
You try to reach the Heavens
So foolish are your ways
So foolish are your thoughts
Do you think you can search the Heavens
Do you think you can reach the Heavens
I am there—I am here
I am in all
Is it not I that set the stars in place
Is it not I that placed the sun and moon
Why do you search the skies
Your thoughts are not My thoughts
Your mind is not My mind
I created all
I will move upon the lands
I will bring down those that should be brought down
I will raise up a people—My people
They will hear My voice
They will seek My face
I will have a people that will stand by
I will do works in them
I will do mighty works through them
Look and see

Wait and listen
For the times are coming
Times of a bringing in—and times of casting out
Watch
Is anything too difficult for Me
Can you not come to Me for your answers
No—you search the lands
You search the books
Why
Can you not see Me
I want to hold you in My arms
I want to shelter you from harm
I want to protect you from all
I want to love you as you've never been loved
My love is forever
My love is great
I created you for My love
Can you not see
Can you not feel
I yearn to hold you—to love you
But I want your love in return
I pure love
A true love
A love that withstands all
A love that lasts forever
For this I have waited
I have waited from the beginning
I know who seeks me
I know the sincerity of your heart
I am looking for a pure heart
An honest heart
One that has a true love
A love that is pure
Have I not gathered you in under My wings
Pulled you in from the outside
Yet you wander out
Stretch your wings
Try your world

But without my protection—death awaits
It will be swift
Come, come to Me
I await your return
I await to gather you—to protect you
Come back, my child
Come back
My Spirit dwells with you
Can you not feel My heartbeat
Can you not know My desire for you
I have wept for you
I long to put My arms around you
Bring you in to Me
The days are getting short
Come to Me now
Come to me

February 10, 2004

I Seek

Oh, how I seek Your face in the quietness of my heart
Oh, how I seek Your face
Oh, what a privilege to know You
To feel Your presence
Oh, for more faith, Father
That I may seek You more

March 14, 2004

I Will

I am going to rip the land apart
I will separate the lands from the seas
The mountains will crumble
The skies will come apart
The winds will cease
The sun and moon will alter their motion
Everything I have created will come to a stop
The stars will cry out
No sound will come from them
I will come with a vengeance
All life will fear
I am coming with force
The stars of the heavens will cry out
The moon and the sun will stop in wonder
The winds will carry My wrath
I will be as a whirlwind, crashing and showing My devastation
All life will halt in their tracks
Nothing will stand in my way
No man can alter what I am about to do
As in the days of Noah
So shall be it
You disgust me to the utmost
My nostrils fill with your filth
I am coming with venom
Be prepared all ye that stand
Be prepared

January 30, 2004

Imagined

Oh, what has man imagined and not brought about
Oh, the lusts of man
Oh, what he lusts after
Oh, does he not seek after that which does not belong to him
Oh, how he searches for which he seeks
And finds it, and makes claim to it
Oh, what punishments come with his prize
What does he fork over for that which he must have
Oh, the lusts of man
Brings about a loss with your gain
Brings about a loss for you, young man
Oh, you look upon another's prize
And you go to all extremes to obtain
That which does not belong to you
Oh, the lusts of the heart
Could bring about the loss of one's soul
Oh, seek after that which does not rot
Oh, seek after that which is eternal
Oh, what you seek after
Be sure of what you seek
Whether it be of God
Or evil
Oh, decide which it is you seek after
For there will be a price to pay
A gain with a loss
Or a loss with a gain

May 15, 2004

Inquire

Where does one go to inquire of the Lord
Where does one go
Is the answer to be found within Your temple
Or can it be found in Your word

March 8, 2004

Insecurity

Oh, insecurity, where did you stem from
Blossoming as you go
Oh, stretching forth your wings
Spreading yourself like wildfire
Taking hold and controlling the ground you lie in
Oh, taking control
And giving your orders
To the one where you have made your bed
Oh, how low esteem takes hold
And insecurity takes the helm
Riding with it, is fear
And unstableness
Oh, how it affects one
Affecting their livelihood
Their family
Oh, how you play your role
Dragging down the one in which you live
Dragging them down to your level
Dragging them down to nothing
Oh, you spirit of insecurity
I speak to you
Remove yourself from this vessel in which you live
For you do not belong here
Remove yourself, I say
For this vessel is not yours to take hold of
You are not to remain here
I say
For you are of the evil one
And this child is Mine
This child is Mine
Take your hands off this child of Mine
I say

May 15, 2004

Joseph's Coat

Oh, what colors were woven into Joseph's coat
The many colors that made the coat most unusual
Oh, until he was stripped of his coat
Until he was stripped, and lay within the pit
Did he become a useful instrument for his God
Oh, how many colors were woven into that coat
Oh, until we get stripped of our coat of many colors
Will we too become useful to our God
His God
The God of his forefathers
Our God
Oh, how many colors are woven into our coat
Oh, is the color of pride amongst the threads
Giving us a prideful walk
Oh, is the color of anger woven into our coat
Or, is theft woven into our coat
Jealously
Lying
Lust
Oh, what is woven into our coat of many colors
Will we miss our most beautiful coat
We too must lie naked and cry unto our God
Oh, what colors do we harbor
That has been woven and interwoven over the years
Oh, how difficult will it be to lay aside our coat
As Joseph did, lying without all that he loved
Left with no one to cry out to except his God
Oh, until we lie within our closet
Stripped naked of all our colors
Will we be useful to our God
As Joseph was
Oh, the woven cloth
Upright threads, but having no strength on its own
The need for the cross threads

To give it strength, stability
Oh, but what colors have we used
Oh, what colors have we chosen
We, the decider of the design
The decider of the colors
Oh, how God provided the upright threads
And how He has set aside the cross threads
He lays them out for us to choose
We, the one to choose which colors to use
Oh, until we lay aside our coat
Lie naked before our God
Stripped of all outside attire
All inside nature
Will we become a useful instrument to our God
We must lay aside our coat
Lay it aside
Pick up the robe of righteousness
Purity
Cleanliness
All the qualitites of a chosen child of God
Will we be a ruler upon the earth
A light unto the world around us
But we must remove our coat
As Joseph did

May 1, 2004

Judgement Chair

Oh, see how I sat on the judgement chair
See how I sat
I judged you who curses
I judged you who robbed
I judged you who murdered
I judged you who lies
I judged you who committed adultery
I judged you who gossips
How I sat on the judgement chair
Seeing all the sins of the ones before me
Oh, how I judged
I saw the sins of others
While turning my back on my own sins
Oh, do you sit in the judgement chair
Judging those before you
For you too will be judged

March 9, 2004

Judgement Chair

Oh, how fast we are to take our seat on the judgement chair
We—who do no wrong
We take our place upon the chair—judging
Judging the
Murderer
Robber
Rapist
Adulterer
Sexual deviant
Oh, how we take our place and sit in judgement
But, what of our own sins
What of the sins we do
Oh, is there a number from one to ten
One being the least
Ten being the worst
Oh, where do we stand in the numbers
Oh, are we a one, or a seven, or a ten
Oh, where do we stand
Oh, is there a number with Me that places the sins of My people into a slot
Oh, they are all the same with Me
Whether a liar or a sexual deviant
Or a murder or a robber
There is not a number with me
For a sin is a sin
Oh, are they not all the same sins with Me
Who can sit on the judgement chair
Who can step forth and take the seat and sit there judging
For there is none innocent
None
None
Who can step forth and say they have not sinned
Who
Who
For there is only One who can say no sin lies within Me

Only One
Only One can step forth
For with Him no sin lies within Him
Yet He took the sins upon Himself
Only upon Him, He took the sins
They did not enter into Him
Only upon Him
Our sins He took, they were laid upon Him
They did not enter into Him
He, who took our sins, did He judge you
Did He place your sins into a slot of judgement
Oh, ye who judges
Oh, don't sit in judgement of others
For your measure of judging
Surely you too will be judged
Oh, beware of who and how you judge
Oh, beware
For the measuring stick is measureless
Measureless is the stick
Oh, did He, who took your sin, did He judge you
Did he take out His measuring stick and measure your sins
Where do your sins measure
Are they a one or a ten
Oh, you foolish one
Be wary of your judging
For you to will be judged
And judged with your own measuring tool

March 26, 2004

Latter Rains

Oh, how You are moving in these days
Oh Lord
How You are moving
Moving across the lands
Moving upon Your people
As the evil one knows his time draws near
He is ever increasing his evil upon the peoples of the earth
Bringing more and more evil
Evil he is spreading like wildfire
Like a forest burning, he has set on fire
It rages and pulls in
Oh, how he mimics
Oh Father, as You spread out Your fire
So too is evil setting on fire
Drawing in anything in its path
Growing in power
Growing in might
Oh, how evil is spreading
Spreading itself around the world
Compounding it self it is
But
Oh Father, as mighty as evil gets
Oh Father, You are in control
You, oh God
Are in control
There is no power
No might like You
For Lord
You will bring upon the rains
You will pour out Your showers
Your heavy rains
They will quench the fire of the evil one
Your rains will pour out
Dampen and lessen the power of the evil

Not to just dampen, but
Oh Lord
Your rains will put his evil out
Oh Lord, how You are bringing Your rains
Oh Father, how that will put a wedge into evil
A blockage
Like a blanket that will snuff out the evil
Oh Father
You saw ahead
You knew of these days
Oh Father
How You saw ahead
Oh Father, you have opened up the doors
Opened the gates
Opened they are
The rains are coming
The rains are pouring
Flooding they are
Oh, how the rain is pouring
Pouring down on Your people
Oh, how it is pouring
Oh how it is raining
Oh Father
Let it rain
Let it pour
Flood us in Your waters
Flood us
Cover us in Your rain
So that the evil one cannot put our fire out
Oh Father
How You protect us
How wet we are
How evil cannot touch us unless we dry out
Oh Father, keep us wet
Keep us saturated in Your rain
Never let us dry out, oh Father
Never let us dry out

March 28, 2004

Lie

Oh, never lie to the Lord
Never
Oh, don't make a vow to the Lord, that you won't keep
Never
Oh, never make a promise to the Lord, that you won't keep
Never
For you don't know what tomorrow brings
Your very next breath
You may be snuffed out before you can fulfill your promise
Your vow
And what of your lying to the Lord
Do you not think He knows you lie
Did He not snuff out the lives of two who lied, whilst they stood on their feet
They lied about their earnings
Did He not snuff them out
Did He not
Oh, be wary

March 14, 2004

Light

What is held within the light
And what is held within the darkness
What lies within the spiritual realm of the light
What lies within the spiritual realm of the dark
Can evil exist with the light
Can good exist with the dark
Is there a dividing line between the two
Is one stopped from crossing into the other
Or do both occupy the light and the dark
What does the evil do when it is overcome with the light
What does the good do when it overcomes the darkness
Which is the strongest of the two
Is not our answer found in the Word

March 10, 2004

Little Faith

Oh, ye of little faith
Oh, how little
Oh, the gifts you seek
The gifts
Are the gifts not given
Have I not given them to you
Yet, you still walk with little faith
Little faith to exercise your gifts
Oh, what must I do to open your eyes
Open your heart
My gifts I have given
Given in abundance
Yet you still seek them
Refuse to walk in them
Refuse to exercise your rightful gift
Oh, ye of little faith
Did I not send out servants into battle
With nothing but faith
Did they not walk in faith
Oh, and what of those that healed the sick
Oh, what of those
Those, who did not have the abundance of gifts I have given you
Yet, you continue to walk around desiring such
Oh, I have given the gifts to all who ask
But where is their faith
Where is your faith
Faith to step out in these gifts
Oh, where is your faith
You, My child, can do so much
Can produce such wonders
And yet, you continue to wait
To still seek gifts
Oh ye are foolish, My child
Where is your faith

April 14, 2004

Locusts

Oh, locusts spreading themselves
Upon the earth
Oh, the grain of the field
Will be devoured readily
Oh, how they are moving across the land
Beware ye wheat of the field
For ye are ripe for harvest
Ripe for harvest
Oh, ye lesser grains
Ye too will be devoured
Oh, the fields are ripe
Ripe for picking
Oh, bring in the wheat
Bring in the grains
Bring them in
Oh, cast your sickle
Swipe over the fields
Gather the grains
Bring them in
For they have ripened
Bring them in
Oh, see the locusts as they move across the lands
Oh, the amount
The darkness that comes with them
Vast amounts
Oh, bring in the wheat
Bring in all the lesser grains
Bring them all in
Leave none for the gleaner
Gather them in before they are devoured
Bring them in

April 15, 2004

Lord

Lord, are we not yet sucklings at the breast
Not yet ready to chew our food
Oh, prepare our teeth and gums
So that we will be able to eat whole food
Teach us to eat, Oh Lord

March 7, 2004

Love Me

How much do you love Me
How much
Your walk is not straight before Me
Your walk is not straight
You say you love Me
Do you
How much do you love Me
Where do your thoughts roam
Do they roam to Me
Do they
Oh, how close is your walk with Me
How close
Is your walk distant
And your love
Is your love constant
Steady
Trusting in your provider
Your Father
Is your love firm
Or are you fickle at times
Where do you run
When you encounter trouble
Do you run to Me for advice
Do you even ask for comfort
When you hurt
When your tears wet your pillow
Do you run to Me
So that I may put My arms around you
Oh, how close is your walk
Oh, how often do you really turn your eyes toward Me
Oh, how often do you read My word
Words to you
Words to help you
My words

Life giving words
A guide book
A help book
All your answers for life are found within the pages
Oh, how often do you read My word
Oh, My eyes are constant on you
My eye does not leave you
You are mine
I wish for no harm to come your way
But you wander in and out of our closeness
Oh, how close is your walk
I desire a close walk
For I love you
Wish for no harm to come to you
Have I not provided all you need
All
Yet I find you wandering
At times afar off
I am always here
Always constant
Always loving
Come back, close, My child
Come back close
For the wolf is out to capture
Oh, come back in close
Don't wander out too far
Come back to me
Come back

April 15, 2004

Love the Lord

Oh, where do we place our Lord in our life
Where do we place Him
Do we love him with all our heart
With all of it
Do we love Him with all our minds
All our minds
Do we love Him with all our soul
All our soul
Do we love Him with all our strength
Oh, where do we really place Him
How far down the list of loved ones is He really
How far down the list of "things to do" is He
Is He foremost in our souls
Oh, where have we placed Him
Does He even have a placement
Or do we wipe the dust off once a week
And bring Him out for show
Do we take time to thank Him for our food
For our health
For our very lives
Do we
And what of the children we have been blessed with
Have we given them back to Him
Or have we forgotten that too
Oh, what have we done with our Lord
What have we done
Where have we placed Him within our lives
Are our loved ones placed first
Are our jobs placed first
Our "other" loves, do they come first
Oh, we cannot just bring Him out when we want to
Bring Him out and wear Him upon our sleeves
For all to see
To sing and dance before Him

Just one day a week
And then for an hour or so
Oh, where have we hidden Him all week
Do our friends really know that we "might" know Him
Do they even know Him
Have we ever taken time to tell them
Are we afraid of being laughed at
And what of our family
Do they know Him
Do they know that you "might" just know Him
Or do you hide that as well
Oh, where have we really placed Him within our lives
Where really have we placed Him
Oh, so simple to say we love Him
So simple to say He is our Savior
But do we
Is He
Oh, come out of the closet ye one
Come out
Come out into the open
Come back to your first love
Come back to when you first fell in love
How you thought about Him all day
Every day
Those around you knew that you had "something"
Then, slowly the day of love became hours
The hours became minutes
The minutes into once a week
And then perhaps, not at all
Oh, come back ye one to whom I speak
Come back
You are to love Me with all your heart
All
You are to love Me with all your mind
All
You are to love Me with all your soul
All
You are to love Me with all your strength

All, I require, all
Come back, My child
Come back
Love Me with all your heart, mind, soul and strength
And I will look after all things for you
I will take care of your cares and worries
I will take care of your loved ones
But come to Me with all your heart
All I say
All

April 12, 2004

Lying

Oh, the lies that come out of our mouth
Clearing us of any guilt
Where do our lies go
Do they really clear us of our guilt
They enter the one we lied to
They are digested
Acted on
Or
Expelled
Then what
Do they just disappear
Or do they return to the sender
Will your master excel in your lie
Or will he try another motive
Oh, what cleverness your master has
What a master he is, trying to deceive you in every way
You fool
Do you not see what is going on, the methods that go on to deceive you
Truth is stronger than lying
Where does your master stand now
Does he try again
Or does he crawl back into the hole he came from
Is it not all up to you who holds the key
Oh, what is your decision, young one

March 19, 2004

Mighty One

Oh, Mighty One of Israel
Your words are strong and true
They cut me down from where I stood
And brought me to my knees
Oh, Mighty One of Israel
Your words did hurt and slice
They cut me to my inner heart
Until the tears did flow

March 3, 2004

Mighty Work

Behold, I am about to do a mighty work in you
A mighty work
I say
For My words you will speak forth
My words are planted within your heart
My words to speak forth
In days ahead
For I will do a mighty work through you
My faithful servant
I am about to do a mighty thing
Oh, my faithful servant
Your exploits will be done through Me
Not like ever before
The power you will walk in
I will give you the boldness you require
I will put inside you
A powerful voice
To speak forth My words
Oh, I will bring all things about
What I have planned
I will bring all things about
For you
My faithful servant
I will do a new thing
Oh, in days ahead
You will walk in
My authority
You will walk in My name
You will walk in all boldness
Not fearing your very life
For yes, you too must suffer as those that walked before you
You too will suffer as they did
Oh, I will do a mighty work in you
Through you

Oh, your hands
Healing hands
Oh, your mouth
A powerful voice
Oh, you will walk
In all authority
In all power
In all boldness
Through a mighty work I will do in you
For nothing is impossible for me
Oh, ye the lesser of them all
I will do a mighty work
Oh, prepare for days ahead
For I will bring this about

May 2, 2004

Mirror

Oh, what do I see as I look upon the mirror
What do I see
What peers its eyes back at me
Do I see ugliness
Or do I see beauty
Do I see hate roll from my face
Or do I see love pour out
Do I see a pent-up face
Or do I see a relaxed face
Oh, what do you see when you look upon your reflection
Oh, we are what we see, are we not
What does our mind tell us as we look upon ourselves
What does it tell us
Does it show a haughty one
Or does it show a humble one
What does it tell us
Do we wear the weight of the world upon our faces
Does sickness pour its face back at us
Does hate reflect itself to us
And how about the eyes that peer back
Are our eyes not the light to our souls
Oh, what do we see within the windows
What can be seen
Do others see within our windows
Do they see what secrets lie within
Oh, what do you see
What do they see

March 8, 2004

Motivation

Oh, what is lust and what is love
What motivates lust
And what motivates love
What holds better
Love or lust
What motivates this lust
Insecurity
Loneliness
Desire
What motivates love
Loneliness
Desire
Which is better
Which will last
Lust or love
Lust—for a moment in time
A passing wind coming and going
A feeling for now but leaving
Nothing left to hold onto
And what of love
A desire—lasting
A feeling—lasting
A love remaining through sickness
Through health
Through good times
Through hard times
Happy moments
Sad moments
Through life till death parts you
God blessed
God saved
Of God
And what of lust
Where does it go

How long does it last
The excitement of the moment
Days lasting or perhaps a year or two
Then what
What happens when sickness raises its head
Do you not sneak out the back door
What of bad moments
Where is your stand now
What of the beauty now gone
Where is your lust now
And now, what of death
Where is your stand now you lustful one
Oh—the desires of the flesh
How long does it last
Does lust follow us to the grave where the earthworm devours us
Can lust continue there
Oh, you foolish one
Who enter into such
Oh, how foolish you are to trade lust for love
One is everlasting
One is for the moment
Oh foolish one

March 10, 2004

My Days

Oh, how you planned my days before I was
You planned my days, oh Lord
You saw the end result, Father
You saw the end
You saw the depths of my sins
You saw my high times
You saw my ups and downs
My failures
My passes
Oh, how You saw it all and brought me through
You saw it all and brought me through
Oh, what was so hurtful for me, Lord
You turned to blessings
Oh, what was so utterly disastrous to me
You brought about good
For Lord, You had Your hand on me, from the beginning
You knew the end result of me
You saw it all
You had Your hand, guiding me
To bring about the end result
For such a time as this
For such a time as this
You brought about this end result
For a day as this
You planned and maneuvered
And brought me to where I stand now
Oh Father, You, my Lord
God Almighty
Have brought all this about
You brought me through it all
To where I stand now
Oh Father
Such a change in me
Such a change in me

For, Lord
You are my Lord
You alone do I worship
There are no other gods before me
There are no other gods before me
For You alone do I worship
You alone do I bow down to
Bend my knee to
Oh Father, You alone
Are doing a mighty work in me
Oh Lord, continue to do and work in me
Until I come purified, purified
Before You, my Lord

May 10, 2004

My Father

I have cast my eye upon you
Why do you wrestle
I have placed you where I want you
You shuffle and turn
But I bring you back
I have plans for you
My plans—not yours
Sit still—lie quiet
See what I will do
You think you are too low—not worthy
But it is you I chose
I shall be the One to guide
I shall be the One that decides
I have My plans for you
Await and see
Await and see
The time is coming to see my works—to see what I can do
It is not you—but I in you
See the mighty works
I will not speak words
I will show my words
Mighty words
Who can stand
Who will deny
You are my instrument
You are here for me—not here for you
I have my mind on you
I see you my child—await and see—wait and see

January 5, 2004

My Grave

Oh Lord, what do I take with me as I go to my grave
Do I take pictures of my loved ones
Do I take my favorite attire
Do I take my jewels
Do I take money
Oh, what do I take with me
And what will I do with what I take
Who can I show the pictures to
Who can I strut before with my attire
Who can admire my jewels
Where can I spend my money
Oh, who can come back from the grave and inform me of this
Who
Is there someone that has come back that can tell me
If I go to Heaven can I take all I've taken with me to the grave
If I go to Hell can I take all I've taken with me to the grave
Who can I show
What can I spend my money on
So, what purposes on earth to hold onto
Pictures
Clothing
Jewels
Money
If when it serves no purpose
What purpose is there
Is it not better to put our riches in a safe where nothing can touch it
Death cannot touch it
Nor can hell
Where it will increase beyond imagination
It won't mold
You can't lose it
Oh, foolish one
Which is better

Investing within the world and taking it with you to your grave where the earthworm will surely devour it
Or putting it in an eternal safe

March 10, 2004

My Heart

My heart cries out for you in the quietness of the night
Where can You be found
Are You hidden within the stars
Are You amidst the winds that blow
Can You be found within the depth of the waters
Oh, where can I find You
Who has laid their eye upon You
How hidden You are
Do our cries reach the uttermost parts of the heavens
Does our tears fall upon dry ground
Do you see us amidst Your glory
Who can touch You
Can the winds gather You in
Can the seas contain You
Oh, when your wrath comes
Hide us within Yourself
Don't forget us when you cast Your power on earth
Oh, protect us, Your children
Gather us in to Yourself
Oh, help us, I pray
Oh, save us from Your coming wrath
Cover us from Your anger
Oh, run, My children, run
For I will sweep the earth
I am going to sweep the earth clean
Oh, where can we run from beneath Your hand
Is there anywhere we can run
Is there any hidden place that You cannot find us
Oh run to Me, My children
Run to Me
My arms are open, My love so strong
We have wronged You, oh Lord
We have done what was wrong
Can You forget our doings

Can You forgive us our wrongs
Oh, what sinful people we are, sinful from the day of birth
Can You find one that did not sin
Can You find one
Don't judge us harshly, oh Lord
We run to You, oh Lord,
Cover us
Help us

February 23, 2004

My House

There is dirt within My house
There is dirt
I ask you to clean out My house
Clean it out
Then I will come unto you
There is sin within the gates of My house
Sin dwells there
Confront evil
Confront it
Gather the leaders together
Filter out the evil
Oh, cast sin out from amongst you
Cast it out
For I will not visit you as you wish
I will not come in My power and might
Oh, ye who are the leaders of My house
Cast out that which is evil in your house
Cast it out
Then, I will come in My power and might
Then I will dwell amongst you
For evil sits within your house
Evil sits
Oh, seek it out
Oh, it lies where you expect not
It lies where you expect not
Oh, search the spirits
Search them out
Which are of Me
Which are not
For the deceiver is out to deceive you
To deceive you
Oh, check out the ones within your house
Check them out
For the evil one is out to deceive you
For My elect will nearly be fouled in this day

Oh, beware—ye leader of My house
Beware
For the evil one desires your house
Desires your take
Oh, beware of the evil one
Beware, ye leader of My flock
Oh, check out the spirits
Check out the spirits
For evil lies within your house
It can bring decay within your flock
Oh, leader of My house
Check out the spirits
Oh, listen carefully for what is said
Oh, check out every word if the words are of Me
Oh, check them out for clever is the evil one
Oh clever is he
Oh, watch the wording
Oh, watch the wording
For the evil one lies out there to pull in your flock
Oh, leader of My house
You are placed there to lead My people
You are placed there to lead My people
Oh leader of My house
You are My chosen one for this flock
Oh, be warned
Evil desires to come into your flock
To disrupt
To separate
To break up
Oh, leader of My house
Take a stand
Take a stand
You are my chosen for this flock
You are My chosen
No one else can lead this flock
No one else can lead this flock
You are their leader
Don't allow the evil one to come and disrupt
Oh, check out the spirits

Check them out
For evil lies within your gates
Evil lies within your gates
Oh, be ever watchful
Ever watchful
Oh, check the spirits if they are of Me
Oh, check them out
Ye leader of My house
Check them out
Oh, ye leader of My house
I will come in My power and might
I will come
I will come with and bless you
I will come and bless your people
I will come
I hear your prayers
I hear you
Oh faithful servant, I hear you
I hear you
I hear you
Clear out the evil ones from within your midst
Clear them out
No two powers may exist side by side
No two powers
Choose which one will encompass and dwell within your house
Choose one
I will come
I have heard your cries
I have heard your cries
I will come, my faithful servant
I have heard
I will come
I will bless
I will bring multitudes into your house
I will bring this about
I will bless you, My faithful servant
I will

April 4, 2004

My Little Child

Oh, how I saw you on your way to death
Oh, I saw you
Watched as you walked your walk
Knowing that you were going to die
I, your Father, saw
There was no other way
But to go ahead of you
Stop your destruction
Take your place
Oh, how I loved you
Ran ahead I did
Ran quickly
Caught your death Myself
Oh, how I gathered up your death
Your entrapment
Your sickness
Oh, how I ran ahead
Grabbed it all
To protect you, My child
To stop you from hurting
To stop you from pain
To stop you from sickness
To stop you from death
Oh, how I ran ahead
Took it all
I, your Father
I, who loves you so much
Oh, how I want to protect you, My child
To put My arms around you
Pull you in
Hide your face from death
To gather you into Myself
To wrap My arms around you
To comfort you

To warm you
To shelter you
From all that would attempt to hurt you in any way
Oh, how I ran ahead to protect you, My child
I, your Father
I, who loves you so much
I, who will always run ahead
Always protect you
Oh, how sickness has no rule over you
Oh, how death has no rule over you
For I have stopped it
I have taken it away
Oh, I have removed it from you
Oh, come to your Father
Come, My child
Come
For I will take care of you
You are never too old for a Father to shelter His child
Never too old to run to your Father for help
Never too old to call out to your Father
Never too old
For you will always be My little child
Always be My little child
Always

April 9, 2004

My Lord

Oh, why do we shut my Lord out
Why do we shut Him out
What guilt brought the door to close
What guilt brought that about
Oh, when did the guilt arrive
Oh, where did it arise
Did the guilt come upon us
Or did we open the door and invite it in

March 8, 2004

My Love

Oh, My people
How I long for you
How I long to hold you
Return from your ways
Come to Me
Return to Me
Oh, I see your tears
Do you not know each tear is counted
Do you not know I watch over you as a mother hen watches over her chicks
I know when you fall
I go before you
Yes, I am behind you
I surround you
Oh, how I love you
How I care for you
I have so much to give you
Ask—receive
How much do you love me
Don't allow your love to grow cold
Where does your heart lie
Oh, check yourselves
What do you see
What are your loves
Oh, check yourselves

February 19, 2004

My Mind Occupied

Oh, how my mind gets taken up by the busyness of the day
Oh, Lord, forgive me
But my mind does not wander over to You
My mind so occupied with my own world
With things to do
How I push You back from the front of my mind
How I occupy myself
And not with You
Oh Lord, forgive me for my weaknesses of this world
Forgive me for not keeping You ever foremost in my mind
Oh Lord, how I go about my life
My life
The very life You gave me
And I go about doing my own thing
And I go hours, even a day
Without even turning my mind Your way
Oh Lord, forgive me of my weakness
Forgive me for occupying myself to such an extent
That my mind does not turn Your way
Oh Lord, how You stand there awaiting my mind to turn Your way
For me to look at You
Oh, how often do I hurt You with my own selfishness
Oh Lord, forgive me for I do love You
And do not mean to forget You
To push You back
For I do not intentionally do it
Just the busyness of the day
Oh Lord, where is my "love the Lord with all my mind"
Oh Lord, You are my Lord
My Savior
And my mind does now go to You
And I pray You will be foremost in my mind
I pray

April 25, 2004

My Mind

Oh Lord, are You not on my mind constantly
You prompt me at the most inopportune times
Your voice so quiet it takes all my might to hear
 Through the noise going on about me
Oh, if You can just speak a little louder so that I might hear every word
I fear I may lose a word here and there with the quietness of Your voice
Oh, if You would speak to me louder
I don't want to miss one word You say to me
Oh Lord, how great You are
How wonderful You are
I love you so much
You are my King
My Deliverer
My Redeemer
You are my all

March 12, 2004

My Nakedness

Lord, do you not see me in my nakedness
What beauty is there to see
Your eyes see me as I stand before You
You saw when sin was conceived within my heart
You saw me fall
Did not wickedness desire me
Luring me to places unknown
I turned my back on You, My Lord, and walked the other way
What amount of tears You shed as You watched me go
How hurtful it must have been to see Your child go their way
How long you watched
How many tears You shed while I wasted my life away
I felt your presence throughout my sins
Oh, Lord, You waited so patiently
My wasted years are now gone
My eyes are opened
I returned to You, My Lord
I felt Your coldness to me at first
But then Your arms opened up
Drew me to Yourself
Pulled me in
Washed me
Clothed me in the finest of clothing
You see my heart
You know my mind
Oh, Father, I was that lost sheep and now am found

March 6, 2004

My People

The sins of My people pile up
You think no one would know your sins
Your neighbour does not
You think they are hidden
But I see your sins
I know your doings
Step out, ye sinful one
Step out
Cleanse yourselves
Your time is nearing an end
Cleanse yourselves
Oh Lord, you cry
I have need of this
I have need of that
Have I not given you the clothes you wear
Have I not given you the food you eat
Have I not given you shelter
Yet you cry out continually
I hear your cries
I know your needs
You are my child
Does not a father care for his child
Would a father see his child go without
Your neck stretches to see your neighbor's riches
You jealous, rebellious people
How long shall I tolerate you
Oh, what do your eyes see
What do your ears hear
Where do your thoughts run
Who will be standing with their clothes washed white
What will be in your thoughts when I come
Oh, cleanse yourselves
For times are coming
Times like never before

Times of sorrow
Times of scarcity
Diseases will run rampant
Fear will be everywhere
Hard times are coming
Will you be ready
Will you be hidden in the shelter
Oh, check yourselves
Wipe your slate clean
There are days ahead
Dark days
Days when the old will be cast out
Days of hoarding
Blood will be shed for food
No light will shine
No warmth to warm you
I am going to sift my people
Your hardened parts will I remove
Oh, you will bend your knee
Yes, all of creation will bend their knee

February 16, 2004

My Sins

Oh, Lord, my sins were so great before You
So great were my sins
Oh Father, what sins were upon me
What dirt covered me
Daily I sinned, did I not
Oh Father, how could You have seen me through my dirt
How could you have seen me
What did You see in me, oh Lord
What did You see in me
I was dirty before You, oh God
Unrecognizable I was
Yet You saw through all the dirt
Pulled me up
Saw in me something, oh Lord
Something that no one else saw
You saw something sparkle within the mud
Like a small diamond hidden in the earth amongst the rocks
You saw that little sparkle
You picked it up this little diamond
Took the dirt off
Took a grindstone to it
Ground off the edges
Smoothed this little diamond
Polished it You did
And kept polishing it till the glitter shone through
And there I was, a little diamond, a little prize
A little prize before the Lord
All washed
All polished
All smoothed
And here I am before You, oh Lord
Already to be placed within your jewels
Already I am

March 15, 2004

My Son

My Son, My Son
My only Son
You were the only One
The only One
In all the heavens
In all the earth
The only One
Pure, pure
White as snow
The only One
That could step up
From the beginning of the beginning
You were there
You knew the end
You knew the pain
You saw the end
Before the beginning
You saw down the tunnel of time
Saw the place You took
You saw the lives lost
The lost over the years
You saw from the first fall
To the last fall
You saw them all
Each and every one
They all hid their nakedness
All hid with shame
You knew before they fell
You knew they would fall
You, saw before man was created
You saw their fall
But, you took Your place
Your place amongst man
Stood in the gap

Made a way for fallen man
Made a way for all
I have given You them all, My Son
I have given you them all
You took their nakedness
You took it all
All are yours, My Son
All are yours
Not one will be snatched from your hand
Not one
For all are born into sin
Born into sin from the first sin
All guilty
All dead
All lost
My Son, you alone have pulled them up
You alone have taken their sin
You alone have brought them in
You, My son
My only Son
My Son
My Son

April 11, 2004

My Spirit

My spirit is moving over the lands
Looking upon the kings of the lands
Looking upon the rulers of the lands
Looking into the very hearts of these rulers
What evil thoughts run wild within the rulers
What evil are they thinking to bring upon a people
Oh, what evils
Oh, the desire to rule all
Gain all
Oh, what evil runs wild within their minds
What evil are they going to bring upon a people
An innocent people, a poor people
A people that are downtrodden
Oh, how evil one can be
The desire within their ruler is about to cause such grief upon a people
Oh, who rules these rulers
Telling them what to do, who to conquer
Oh, who tells them when to move, when to sit back
Oh, how they listen to the evil one, blinded they are
Blinded by the master of evil
The master of deceivement
Oh, how they listen to him
Oh, ye ruler of the lands
Ye kings of the lands
Open your eyes
Open your heart
See who really is ruling
For you are nothing but a mere puppet in the hands of the evil one
A mere puppet
And a people will suffer for your weakness to follow after the evil one
They will suffer
An innocent people
A poor people

Oh, ye ruler
What are you about to do

March 23, 2004

My Thoughts

Oh, what thoughts get implanted within my mind
Where do they come from
These thoughts
So, carefully implanted there
They just pop up within my mind
Out of nowhere
Oh, thoughts of You, oh Lord
Your blessings
Your Love
Oh, Lord thoughts of You
But—then, oh Lord
I get other thoughts not of You, Lord
Thoughts not my own
Thoughts I could not even dream of
Yet, I find they sneak in
Slide in the back door of my mind
Implant themselves there
Waiting for me to take notice and make my move
Yet, You also, Lord
Await my move
Do You not
You wait my move with Your implantations
You wait my move for the other thoughts that race through my mind
Oh Lord, You stand back
Awaiting my action
Am I going to take action with my thoughts
Whatever they may be
Am I going to step forth and act upon them
Oh Lord
Keep me from the evil one
Plant a hedge around me
Keep my mind clean
Keep it holy unto You, Oh Lord
I need Your help

For I cannot do it without You, Oh Lord
Oh, forgive us, for we are weak, are we not
Allowing thoughts to cause us to take action and fall into sin
Oh Father, give us the strength to cast that out that is not You
Give us faith to stand for You and not fall into sin
I pray

March 14, 2004

My Twilight Years

Oh Lord, You waited till my twilight years
Before You came to me
In Your abundance
You waited through my youthful days
My rebellious years
You waited through it all till I matured
And able to accept all You have for me
You waited
Oh Father, how foolish I was to throw the many years away
How foolish I was

March 14, 2004

My Word Goes Out

Oh, hear My people
My word goes out
Incline your ear
Listen carefully
Take My word more swiftly
Go fast
Go forth, take My word to a stubborn people
Go swiftly—go
My word will pierce the skies
Did not My word create all
Did not My word do all things
Oh, you foolish man
Why do you question My word
My word is true
Oh, check and see
Measure the length
Measure the breadth
See for yourself
Does not My word hold the sun and the moon, the stars in the heavens
Oh, who knows where My word goes
What ear it reaches
Who can judge
Which ear receives
Which ear does not
Who has seen the power
Who has seen the tenderness
Who can judge who accepts
Who can judge who does not
Oh, Lord, your word has touched my inner being
My inner parts cry out to you
Oh, how mighty are your words
Does not all creation sing out to you
No, not all creation you say
Oh, tremble whose ears My word does not tickle

Oh, tremble whose ears does not hear
My word has gone out before the beginning of time
My words will remain forever
Oh, take note
Listen to My words
My words are true
They will remain forever

February 22, 2004

My Word

My word did not come from one you would expect
My word did come to them
I gave My word to one unknown
From them you will hear My word
Was your Lord born from royalty
Or did He come from an unknown
Oh, hear the words from where they come
Oh hear, for the words are Mine

March 9, 2004

Mysteries of the Body

Can one know the mysteries of the brain
Can one know the mysteries of the eye
Can one know the mysteries of the ear
And what of the heart
Lungs
All internal organs
Oh, what master designer designed all this
And what of the circulatory system
And all the other systems working in one's body
Who drew the plans for all this
Who can step forth and design one as great
Who was the master planner of it all
How intricate our parts are
One working in unison with the other
And what of the wonder of our reproductive system
To be able to reproduce such a being
Oh, the wonder of it all
Has anyone been conceived that can compare to bring forth such a plan
Our minds can't comprehend the depth of knowledge this Master Designer
has
No, there is not one
He alone was able
With His word

March 6, 2004

Mysteries

Oh, what mysteries hide within the mind
What brings about the thought process
And where does the thought originate
Then where does it go
Does it travel its course to the heart
Where does it go from there
Is it eliminated from the body
Or is spoken out the mouth
And you, oh tongue—what have you to do
What words have you spoken today
Did you gather your words from within the heart
What will these words produce
Will they bring about life
Or will they bring about death
And what of death
What brings that about
What motivates it
Oh, what happens to one that dies
Where do they go
Where do they go
Oh, the mysteries of the mind

March 8, 2004

Nations

Oh, nations crumbling
Flags dismantled
Oh, buildings falling
Cities destroyed
People running
Cries coming from their lips
Oh, what is happening
All around
Where to run
Where to hide
Watched where ere we go
Oh, constantly watched
Where to hide
Where to run
Watched everywhere
Eyes hidden here and there
Who can we trust
Who can we run to
What country is safe
Oh, where to go
Oh, run to the Father
Run to Him
Let Him hide you in the secret place
Hidden from the eyes
Hidden from those that seek you out
Oh, run to Him
Hide within Him
For He will enclose you from the eyes
Run to safety, My child

March 22, 2004

New Things

Oh Lord, are You not about to bring forth new things
Oh Father, how great You are
Oh, what are You about to bring forth
Father what
Oh Father, what goodness You are showing Your children
But, oh Father, what else are You bringing forth
Oh, to those not called the children of God
Oh, woe to you
Woe to you
Oh, sit and write
Write the words I give you
Sit and write
Words of Mine
These words will go out
They will reach a people who are
Hurting
Bleeding
They walk in the dusk, only because of their hurting
They have a veil before their eyes stopping them from seeing behind it
These words will mend their hurts
Seal over the bleeding
Remove their veil
And they will once again walk in the light
Oh, sit and write these words
Sit and write
Oh Father, what of those not called the children of God
Are we not all Your children
Oh Father, what of those
They are not My children
They are the children of the evil one

March 17, 2004

New Wine

Oh, as new wine cannot be poured into old skins
As the bubbling up of the new wine will cause it to burst
So is with My spirit
It too cannot be poured into old skins
New skins are required
For the bubbling up daily
Will not be able to hold the new wine
Ye must be born again
New skins, tender, yet firm
Able to stretch
Able to contain
Oh, the old skins
Brittle they are
Crack and tear easily
Cannot grow as the new wine matures
But the new skin
Will grow with the wine
Able to hold it
Able to mature with the new wine
Oh, fresh, new skin
For My Spirit
I say

March 26, 2004

Oh Father

Oh, Father, how You look inside of me
You look where no one else can
How You know us like no other
How You know our thoughts
Our every thought
Oh, how You know us
Do You not look through the powers and see our every move
Oh, Father, You know us like no other
Oh, Father, how great and powerful and wonderful You are

March 14, 2004

Oh Call

Oh, call the peoples of the lands
Call them out
Bring the musicians
Prepare a song
Be ready to clap your hands
Dance a dance
Oh, clear your voices
Prepare to play
Oh, bring My people from the north, from the south, from the west and from
the east
Oh, a praise is coming
A glorious praise
Prepare your instruments
Oh, check your feet
Be ready for a feast
Oh a joyous time to come
Prepare to play your instruments
Prepare to clap
Prepare to dance
Prepare to sing
Oh, a time is coming
Prepare for the King

March 1, 2004

Oh Evil One

Oh, where have you walked today
Oh evil one
Where have you walked
What evil did you do this day
What lie did you produce
Oh, where have you walked this day
Evil one
What deceivement did you enhance
Oh evil one
Have you not lost
You so clever
And beautiful in your walk
Did you not sing amongst the angels
Did you not inhabit the greater heavens
Did you not have a high standing place
Oh evil one
You have been cast out of your eternal home
Oh—where do you roam now
Oh, what mischief do you create now
You pulled yourself down—and brought many down with you
Oh—where is your beauty now, oh evil one
Where is your mighty walk now
Do you not crawl within the earth
Poking your head into whatever vessel that allows you entrance
Oh evil one
What have you done
Is there any regret within yourself
Is there any remorse for what you've done
You thought you could sit within the highest
What gave you the right to think such a thing
You are cast down
You were happy to be cast down
A kingdom of your own
Oh, you have now been put down again

Your kingdom taken away
Where is your might now—oh evil one
Where is your power now
Oh evil one
You are but nothing
Your beauty gone
Where are your jewels now
Oh evil one
Where is your strutful walk
Your power now gone
Oh, where do you stand now
Oh evil one
Where

March 7, 2004

Oh Father

Oh, Father, what a wanton selfish people we are
Does not your love expand the universe
Do you not hold all in place
You set the stars in the skies
Do they not sing out to You
You see the trees bow down
Your angels surround You
Oh, Lord, we are so small before You
Do not the birds sing out to You
Do not the winds obey You
Do not the heavens declare Your glory
Oh, Father, You care for the birds that fly
How much more is man to You
Can the birds declare Your glory as man is able to
The animals that roam and the fish of the seas, can they bow down and declare
You their king
Oh, Father, you have created us to worship You
Oh, hear the cries of those now lost
You know the end from the beginning
You know our ways
Father, gather those now lost
Bring them home
Oh, have compassion on them
Allow us time to find the lost
Search them out
Oh, Lord, You were there when earth was unveiled as it poked its head within
the waters
You saw the green grass appear
You watched as the flowers blossomed
The trees gave life
You were there at the birth of earth
Oh, God
Did you not say that if you found one righteous that you would not destroy the
land of old

Surely, now, there must be one righteous
Surely one
Are we not cleansed from sin
Are we not now counted as righteous
Oh, can you not hold back your wrath for us Your children
Oh, watch what I am about to do You say
Listen to My warnings
See the skies—look for the signs
Then you will know of what I speak
Oh, watch and see

February 25, 2004

Oh Happy Man

Oh, happy is the man who has found his way home
Oh, happy is this man

Oh, happy is the man that finds pure love
For he will be well looked after

Oh happy is the man that no evil substance touches his flesh
For he will retain his knowledge

Oh, happy is the man that works hard
For he will be rewarded

Oh, happy is the man that saves his earnings
For he has saved wisely

Oh, happy is the man who provides well for his family
For his family will remain a family

Oh happy is the man that rears his children right
For he will be respected

Oh, happy is the man that disciplines his young ones early
For they will show love

Oh, happy is the woman that finds a good man
For she has been led by God

March 6, 2004

Oh Holy Spirit

Oh Holy Spirit
You came to me in the depth of my sins and drew me towards my Lord
I think back now and recall the many times You fluttered around me
Oh, how blind and deaf I was
How foolish I was to have passed You by
How fortunate I am that You came to me and introduced me to My Lord and
Saviour
He was always there standing in the back ground
Was He not a true gentleman, never forceful, but always there
Protecting and loving
Oh, how fortunate I am to open my eyes and see
How foolish I was to have walked my own walk
How utterly lonely without You
I thought I could go my life alone, learning, acquiring, loving
Oh, how foolish I was, how meaningless my life was
Oh, Holy Spirit—how gentle You are
Father, I thank you for planning so far ahead, for choosing me when I was yet
a twinkle in my fathers' eye
You chose me yet knowing the times I would fall
Holy Spirit, thank you for wooing me to the only One who could wash me
clean
The only One that could pull me out of the mud
Oh, I stood in quick sand, sinking continually
You caught me and pulled me out
Washed me clean
Lord Jesus—You did so much for me
You removed my blinders, unplugged my ears, renewed my life
Gave me a hew heart
Oh, how fortunate I am, how fortunate I am
Lord, You were always there
Were You not watching me, standing in the background, picking me up every
time I fell
A child I was, learning to walk, You were there pulling me away from the many
accidents I could have had

Oh, how foolish I have been to have walked in sin while You stood back, still waiting, but gently tugging me to Yourself

Oh, how You stood as I left my toddler years, and entered the years as a rebellious youth

You stood by and my adult years were a steady coming and going

Busy, no time for listening for Your voice, no time to feel Your gentle nudge

Now as my eyes are growing dim, I see more clearly now

I have time to feel Your presence

Hear Your voice

Oh, how fortunate I am for the years You gave me to grow into adulthood

Oh, how thankful I am

February 27, 2004

Oh Hush

Hush, ye people of the earth
The sounds of hoof beats afar off
Woe to earth
Woe to earth

February 27, 2004

Oh Jesus

Oh, Jesus, how can I ever repay You for what You have done
Giving me an entrance, an opening into the very throne room of God
Oh, not just to stand on the outside looking in
Not just to stand in the vestibule
But to open the door—You, Lord Jesus
And open it wide
And allow me access to go in and bow down before my God
To sing before Him
To dance before Him
Oh, Lord Jesus
There is no way to repay You
No way
Oh, how it was planned before the foundation of the earth
Knowing man would fall
Us weaken souls
You made a door
A way
A veil
A veil that is split in two
Allowing us in
If we locate the veil
For the way there is narrow
Oh, so narrow
Happy is the man that locates it
Oh, happy is this man
For he will go in and out to pasture
Oh, happy is this man

March 14, 2004

Oh Lord

Oh Lord, was there anything new to You as You stepped down
From Your throne and entered upon earth
Was there anything new
Oh Lord, did You encounter our weaknesses
Our struggles
Oh Lord, You stepped down to save us
And oh Lord, did You not see and encounter the weaknesses of man
Did You not see the struggles we have, oh Lord
We can never come to Your level
No matter how hard we try
Not without help
Oh Lord, help us in our struggles
For Lord, we need more faith
We need more You
We need more power
Oh Lord, help us as we travel through this life
For You know
Oh Lord help us
For You encountered all our struggles
All our weaknesses
All our sickness
Oh Lord, help us who are weak in faith
Help us

March 10, 2004

Oh Lord I Cry

Oh Lord I cry
We are so blind
We are so deaf
Oh, peel away the blindness
Remove the plug within our ears
Oh, soften our hearts
Warm our blood
Oh, search us
Oh, cleanse us
Hear our cries
Oh, search the earth
Look for the pure in heart
Seek out the pure in mind
Oh, search the earth for us
Our hearts are pure
Our minds now clean
Oh, how washed we are
Oh, how clean we've become
We were so lost and now found
Blind now see
Deaf now hear
Love replaced hate
Sinful, now sinless
Oh, how great is Your love
How great You are

February 23, 2004

Oh Lord, Mighty You Are

Oh, how mighty is my God
How powerful You are
Your beauty I see
Your love I feel
How strong Your hands
You hold all in all within Your hand
Your love so powerful
You dance amidst the stars
You speak and a thousand words come out at once
Oh, how lovely you are
Your riches overwhelm me
Your love makes one tremble
Who can return such a love
Who can equal it
Oh, how unworthy we are
You paved a way to match our love
Oh, how great You are
Your words are as a mighty wave continually replenishing,
continually loving
Oh, the stars sing out
The trees dance before You
Oh, Lord, my God
You have made me worthy
You have counted me in
I dance before You
I cry before You
Oh, Lord, don't be too harsh on us, your children
Oh, Lord, consider all you've made
Oh, see within our hearts
Oh, there must be some good to be found
Oh, mold us to your liking
Oh, shape us as you must
Oh, remove that which is decayed before you
Renew our parts

Slow your anger
How long shall I hold My anger you say
How long must I look upon the earth
Your sins run over
Is there nothing man has not done
I made a way, who has taken that way
Who has fallen on that way
Oh, the path is there
But you stubborn people, you enjoy your sins, you look the other way
How long shall I put up with you
How long
Oh, Lord I cry—hold off your anger
So many will fall
So many so lost
So many so sinful
Oh, Lord, widen your path
Open up the doors
Oh, help this sinful world
Oh, help us

February 23, 2004

Oh Lord, Remember

Oh Lord, when You raise us up on the last day
Oh Father, remember my family
Remember my prayers
Oh Father, don't let my prayers go unanswered
Oh, remember my prayers and raise them up too, to be with You
Oh father
Remember my prayers
And oh Father, what of those that are not covered with a blanket of prayer
Oh Father consider them
Oh Father, I pray for them too
I pray for all of them, that they too will be raised up to be with You
Oh Father, hear my prayers
Oh Father
Remember me on the last day, and recall my prayers I ask

March 19, 2004

Oh My Child

Oh, My child
Are you lost out to sea
Lost in the waters without your boat
Oh, My child, don't be lost out to sea
Struggling within the waters
For the shark is circling the waters
Looking for those lost within the seas
Looking for the strugglers
Oh, My child
Get in the boat
For the shark is approaching and will pull you
Into the depths of the waters
And into darkness
Oh, get in the boat

March 15, 2004

Oh Proclaim

Oh, proclaim the name of the Lord throughout the lands
Proclaim His Holy name
He is the Lord of Lords
The Kings of Kings
The Prince of Peace
Deliverer
Healer
Savior
Provider
Master
Wonderful
Holy
Oh, proclaim His Holy name
All ye peoples of the earth
Proclaim His name

March 2, 2004

Oh Satan

Oh Satan, you go about the earth like a roaring lion
Lying in wait to attack
But you come not as a lion, but as a lamb
A lion within sheep's clothing
Oh, how deceitful you are
The ability to deceive
How we must guard ourselves continually
Encampeth within the circles of the Almighty
Be surrounded completely within Him
Don't even have an opening anywhere about you that he can slide in
Oh, continually search yourself
Continually check your doors
Your windows
Check for cracks
Oh, be warned
He does not come as the roaring lion you expect
He does not come in ugliness as you expect
He doesn't come in his power and roar
But
He comes gentle as a lamb
All cute and cuddly
So lovable
Oh, be warned
Be warned of the lion within the clothing of a lamb
Do not be deceived
Do not open up to him
He awaits to devour you
He awaits
He awaits
Oh, be warned
Put on your protective clothing
Gird yourself within your clothing from the tip of your head to the bottom
of your feet
Oh, wrap yourself in your clothing

Don't allow yourself to be caught with your guard down
Oh, be warned
This is not a lamb
This is not a lamb
But this is a lion
A roaring lion
Who is circling the earth with his anger
Oh, children, don't let your guard down
Keep yourself tight within your house
Keep your doors and windows closed and locked
Oh, remain inside and keep your sword with you
Don't go anywhere without your sword
Nowhere without it
For you have been warned

March 12, 2004

Oh, the Lonely

Oh, the lonely, my God
How they are ever present before us
Oh Lord, the lonely
Walking alone
Living alone
No one to care
No one to care if they awoke in the morning
No one to care if they even ate
Oh, how they are ever so present around us
Daily we meet them
Daily we see them
Oh, we should reach out to them
For Lord, their cries of loneliness reach Your ears
Their tears You see
Oh, Lord, You are ever present with them
Your face they seek continually
And Lord, You cause them to cross our path
For us to reach out to care
To love
Oh Father, You see the reason for their loneliness
You see it all
You see the losses they have suffered
Losing a loved one
Leaving them so alone
Alone within the world
Without their loved one
It may be a child
It may be a parent
A loved one
That they have lost
And so alone they are
Oh Father, you fill their needs
You fill their void
And Father, you bring those who care

Those who will care for them
To fill that void
To hold them in their arms
And tell them they are loved
That they are so loved
Oh, take time to see around yourself
Take time to see those that walk alone
That shed their tears in secret
Oh, take time to search for them
For they are always around us
Always crossing our path
Oh, draw them in
For God, their Father, will bless you
He will bless them that bless His children
Oh, take time out to pray for them
To show them you care
For they walk in such loneliness
They walk in such loneliness

May 9, 2004

Oh, the Sound

Oh, the sound of thunder in the heavens
A powerful storm approaches
Can you hear the sounds
Can you feel the power of it
Oh, it gives movement to the earth
Does not earth shake beneath its power
Oh, a mighty storm
The earth will quake
The seas up tide
The trees bend over
The mountains crumble
Oh, the sound of cries throughout the lands
Heartfelt sorrow everywhere
Oh, quietness comes
A hush, a stillness
Time stops in its tracks
Oh, where is your might now, you people of the lands
Where is your strength
Where can you draw your strength
Which well would you use to draw from
Oh, fear abounds
These days are coming
Days to fear
Which well did you choose
Where did you get your strength
Has it proved itself
Did you try My well
My strength
Oh, these days I speak of are in the writings
They have been spoken of—warned of
Oh, stop, take note while there is yet time
Seek out the other well
Try it—prove it
Your days may be cut off while you stand on your feet

Oh, I am a loving God
A patient God
A God of compassion
A God of healing
Life giving
But—I am a God of force
A God of anger
A God of hate
How long must I tolerate your coming and going
Your selfishness
Your greed
Your lying
Your cheating
Oh, you cannot be with two faces before Me
Choose one or the other
You have your choice
Don't delay
Don't delay

March 1, 2004

Oh, Watch

Oh, watch what comes out of your mouth
Oh, watch
Oh, what lies within the tongue
Oh, how it can trick you
Oh, watch
For it brings forth what lies within the heart
Exposing you for who you are
Oh, watch
For your tongue can bring forth
Health—and life
Sickness—and death
Happiness—or sadness
Oh, watch your tongue
For it is out to deceive you
Oh, watch

March 14, 2004

Oh What Do You See

Oh what do you see today, My Lord
What does Your eye see
Do You see a people with a humble heart
A worthy heart
A prayerful heart
Does Your eye see as we go to give You praise
Does Your eye see
Are our hearts worthy of You as we go to the house of the Lord
Do we hear Your word spoken
Does your eye see as we bow down in prayer and see our minds drift off to
worldly things
Oh, help us in our weaknesses, Lord
For we are a fickle people are we not
Oh Father, help us in our weaknesses
Help us

March 7, 2004

Oh What Evil

Oh what evil lurks within your heart
Was it conceived within your mind
What action will you take with this evil thought
Will you tuck it away for future use
Will you dig it out at the first opportune time
Will you support the sin
And take the steps it beckons you with
Oh, what will you do, My child
Oh cast it out
Don't entertain it
The road back is difficult
Change directions
Cast it out before you take any steps
Evil is like a web
It encases you
Pulls you in
Drags you into the center of the web
And devours you
Oh, be warned
Be warned

March 6, 2004

Oh Where Have You Been

Oh, where have you been this day, My son
Where have your feet taken you
Have they taken you to the house of sin
Or have they taken you to the mountain of the Lord
Oh, what have you spoken about this day, My son
What curses rolled off your tongue
Or was it praises to your Lord
Oh, what did your eyes look upon this day, My son
Did they look upon the obscene
Or were they cast upon My word
Oh, what did your flesh do today, My son
Did it do a sinful act
Or was it laid aside for the Lord
Oh, what have you done this day, My son
What have you done

March 6, 2004

Oh Where

Oh, where does one go for knowledge, my Lord
Where does one go for wisdom
Where does one go for truth, my Lord
Where does one go for insight

Do you know of where the secrets are hid
Does one know where to seek them
Does anyone teach of where these are to be found
What books does one buy to study the ways

Oh, where can one find the secrets of life itself
Where does one look
If I traveled the world around
Is there anyone that can help me to find

Oh, where can one go to find these precious gems
Is there someone to tell me the way
Oh, where is the way to locate these precious gems
Oh, where is the way

Oh, is knowledge not given to one as a gift
Is not wisdom the same
And what of truth and insight
Are these given as a gift as well

Oh, who can lead me to these gifts
Can one purchase them with their gold
Or are they passed down from generation to generation
Or are they given of the Lord

March 7, 2004

Oh Wicked One

Oh, wicked man
What do your eyes see
Where do they look
Better off if you had been born blind
Oh, wicked man
What do your ears hear
What do they hear
Better off had you been born deaf
Oh, wicked man
What does your tongue say
What words does it produce
Oh, better off had you been born speechless
Oh, wicked man
What does your flesh touch
What does your heart tell you
Oh, better off had you not been born
Oh, happy is the man that sees good things
Oh, happy is the man that hears good things
Oh, happy is the man that his tongue speaks good
Oh, happy is the man that his flesh and heart doth no wicked thing
Oh, happy is this man
For he has life

March 5, 2004

Oh Young One

Oh, young one
What have you done with your life
What have you done
Have you done all you wanted
Gone where you wanted to go
What else do you desire to do
Where else do you desire to go
Oh, are you satisfied with what you've accomplished
Are you satisfied with your loves
Your affairs
Your drunkenness
Your life
Are you truly satisfied
Or do you still yearn for more
For whatever lies across the bridge
Is there more
Will it bring you satisfaction
The desires of your heart
Oh, young one
You wiggle and wrestle
Restless you are
Unsatisfied
You've tried it all
You've tasted of the wine
You've tasted of the lust
You've tasted it all
Yet, you still yearn
Yearn for whatever lies ahead
Whatever will quench your thirst
Oh, if I could only have that
Only have this
Then I will settle down
Be satisfied
Oh, young one

I have placed inside you a yearning only I can give
Only I can fulfil
You can look throughout the earth
Search it high and low
Enter into whatever your desires are
But, you will not find your true satisfaction
Your true rest
You will not
You will not
Until you try Me
Taste of Me
Then you will find your rest
You will have all your desires satisfied
Man and God have a bond
A link
Until that link is connected
You will not find true rest, My son
For I have come to give you life
To give you life in abundance
You will not lack within yourself
You will be filled
Be filled, My son
Be filled
Try Me
Taste of Me
Oh, the honey is sweet
See if it is true
Oh come, My son
See the waters, they will cleanse you
Fulfil you
Oh, come into the waters, My son
Step in
Try me
Do not continue to search around
Oh, restless one
Step in and try Me
For I am pure
I will fulfill all your desires

Satisfaction you will feel
You will enter into peace
My peace
Peace, My son
Peace

April 12, 2004

Oh, Flee

Oh, flee, you children of the evil one—flee
For I am about to stand in the heavens
I will cause the earth to shake off its axis
Darkness will appear
No light to shine upon what I have created
The stars will tumble
The seas will roar
The mountains will crumble
Oh, you children of the evil one
Sorrow lies ahead
I will destroy the evil from earth
No evil will stand before Me
Have I not told of days to come
Have I not warned
Oh, come out of My children, oh evil one
Come out
For I will make all things new
No wicked will be found there
Oh, evil one—you hide within My children
You lie in wait in them
You hide within their bodies, even into the very depths of their bowels
Oh, flee out of them, oh evil one
For I will gather in My children
They will be a pure people
No evil will be found in them
Oh, flee, you evil one
Flee

February 26, 2004

Oh, My People

Oh, My people
I am looking for a people
A people of pure heart
A people of pure mind
They will search My word
They will speak My word
They will bring My word to a people
A people of hardened hearts
Oh, go, My people, tell of My love
Bring them in
Oh, bring them in
I wish for none to be left
No not one
Oh, search My word
Go forth
Speak it out
Bring in a people of cold hearts
Oh, go
Oh, speak
Oh, bring them in
Oh, I am preparing for these people
I have made room for them
Oh, they will come
This people of hardened hearts
They will melt before My words
Oh, they will come
Go forth
Speak My word
Bring them in
Oh, go, My people
Shout My word from the mountain tops
Go to the lowest lands
Speak My word
Go to the north

Go to the south
Go to the west
Go to the east
Oh, bring in a people of hardened hearts
Oh, bring them in
My word they will receive
I am preparing their hearts
I am bending their frigid bodies
Oh, they will come
Oh—go speak My word
Oh, go
Oh, be warned for I am coming
Yes, I am coming
Yes, My words will quicken
My words are sharp
They will cut
They will remain

February 21, 2004

Open Door

Oh, how we open the door for the evil one
He sits awaiting
Awaiting on the sidelines
Awaiting for an open door
Oh, how quickly we fall into anger
Not great anger, but just a hint of it
Oh, how we fall into temptation
Or the little white lie
No harm will come from it
Just a little white lie
Oh, what of the lust we suffer
Oh, just a hidden desire
Only we know of it ourselves
Oh, how we slip
Oh, how the evil one is awaiting
Oh, we open the door
Sometimes we open it wide
Sometimes only a crack
But open is the door
Allowing the evil one to slide in
Oh, how he plants temptation in front of us
To cause us to slip
And slip we do
Oh, where does our anger come from
Our lust
Our temptation
Our white lies
Oh, where do they come from
Oh, how clever the evil one is
To tempt us at our weakest point
He knows our weakness
He has been there before with you
He awaits for your fall, again
Oh, the unseen world

The unseen world
What moves around us
Steady watching us
Steady following behind
Oh, how he resided with us before
All settled in
But now he has been cast out
He sees the light within you
And desires to move back in
Put out the light
Oh, how he maneuvers himself
Oh, clever he is
Oh, children
Watch your every thought
Your every action
Your every spoken word
Think well before taking action
Oh, how clever the evil one is
How he is after all that he lost
Oh, how he is bound to recapture
Oh, watch your every step
Oh, if you see your brother or sister fall back
Oh, run back
Pick them up
Don't criticize
For you too have at times fallen back
Pick them up
Tend to their needs
Oh, how evil wants to slide in
Oh, how difficult to remove him after each entry
Oh, how difficult it is
Oh, don't allow him to blind you
To trick you
For he is after you
You, who he lost
After you he is
Oh, keep your door closed
Keep it closed

Keep your thoughts clean
Don't fall into temptation
Watch your innocent white lies
For they are not innocent
A lie is a lie
Oh, watch your mouth
Guard it carefully
Keep yourself on guard daily
Daily walk in the word
Daily walk in the word
For the word in written within you
Oh, keep it handy
Think on good things only
Keep My word

April 17, 2004

Our Days

Oh, how our days are numbered
The days allotted us
By God Almighty
Oh, does anyone know the day or the hour that they will depart from this
earth
Oh, we all have a date that has been set
An hour, a minute
When God will take us out
Oh, can anyone know the time
Or the date
That He will take us off the face of the earth
Oh, we think we can buy time
We think we can stretch our time here
Oh, can anyone bring about another day here
Oh, can we ask for an extra time here
Can we
Oh, it has been done
Oh, it has been done
Oh, we can stretch our time here
We can lengthen our days
We can cast off the sickness that will take us
We can prolong death
We can
For He has given the strength to do this
He has given us the strength to prolong this death
Oh, we must cast off death
We must cast off this spirit of death
For our Lord reigns
He reigns
He can do all things
We can do all things
With Him
Who lives in us
We can do all things who lives in us

We can raise the dead
We can cast out sickness
For He who lives in me is able
To cast out that which is not of Him
He is able to do all things
For he alone is Lord
He alone is Lord
And I can do all things through Him
Who lives in me
I can do all things
For He is working in me
He is working in me
To bring this about
He can bring this about
I am His vessel
I am His vessel
And I can do all things through Him
For I am His vessel
To be used as He pleases
I am His vessel
I am

May 9, 2004

Our Father in Heaven

Oh, Father
Give us this day our daily bread
Oh Father
Your word
Your bread
Oh Father
Give us daily Your bread
That we might eat
And go out full
Oh, Father
Let Your will be done
Your will
Oh Lord
Your will that I live my life abundantly
That I will be filled with joy
That I walk in perfect health
That I shall not be barren
Oh Lord,
Your will is endless
Endless is Your will
Oh Father
Forgive us our trespasses
As we forgive those who have trespassed against us
Oh, but those who have trespassed upon me, their trespasses are greater than
mine
Oh Father, does not their trespasses over weigh mine
Forgive us our trespasses, Lord
As we forgive them that have trespassed against us
Oh Father
Deliver us from temptation
All our temptations
Deliver us from the evil one
Amen, Oh Lord, Amen

March 11, 2004

Our God

Oh, how You Father, the God of All
The One that walked and talked in the garden
With Your creation
Father, You the One to remove all humans from the earth except for Noah
and his family
And two of all flesh that had breath
Oh, You Father, You the One to establish a covenant with Your people
Oh, You Father, the One that led Moses through the desert with a cloud by
day and a pillar of fire by night
You Father, Who allowed Moses to see the back of You as You passed by him
Oh Father, You the One That destroyed those that had left Egypt to go to the
promised land
You destroyed them
Oh Father, the laws You sent forth for man to walk in
Laws, Father, that no one could keep
Oh Father, You asked for sacrifices to be made for their ever increasing sins
Oh Father, You the One that lead the armies of Your chosen
You fought their battles
Oh Father, You the One that punished Your people
You Father, the God of all
You Father, who no eye would see lest they die
You Father, seeing that man could not adhere to the laws
Oh Father, all mankind would be destroyed again
And Father, You stepped down from Your kingdom
You entered into the flesh of man
Your eyes looked out of a mans eyes
Your life as ours, to be tempted
To feel anger
To feel love
To eat as we eat
To walk as we walk
Oh Father, You became a man
To know our hurts
To feel our pain
To see as we see

Oh Father, You the God of all
Became like us, to be like us Your creation
Oh Father, we still fall into sin
We still cannot abide by the laws
But Father, You became man for that reason
No longer the sacrifices to be made
To cover our sins
No longer will do we have to enter Your tabernacle with fear
Oh Father, we can come right up to You
Sit down and talk with You
One to One
Feel Your protection at all times
No longer must we run and hide from You
For we can come into the Holy of Holies
And talk directly to You
Oh You, our Father
How mighty and powerful
Gentle and loving
Oh, Father
Let us always remember Your position with us
Let us always remember our place with You
To not take You for granted
To hold You in respect at all times
To reverence You in our rightful place
For Father, You are the God of all
And we the clay
We the clay, oh Lord

May 13, 2004

Our Skin

Oh, how we look upon a color not our own
A race not our own
Eyes not our own
Ours to be the superior race
The most intelligent one
Oh, how our eyes look upon the other
The lower in our eyes
Oh, how we judge
Looking down on those that do not appear as we do
Oh, are we not all one flesh
Oh, do we really look upon the lame
The cripple
The blind
The deaf
Do we really treat them as our equal
Does one treat the other as their equal
Really
Do they
Oh, how we show others we do
But do we
Really treat one as our equal
Does the black, really see the white as their equal
One to take into their home and be treated as their brother
Does the white really take the black into their home as their brother
Does the slanted eyes take the un-slanted as their brother
Do they
Do we
Oh, which did God make the superior race
Which one
Did He make the white the one
Did He make the black the one
Did He make the slanted eyes the one
Oh, which one will be the top notch with God
Oh, are we all not one under the skin

The same organs
The same blood
The same all
Are we not all one under the skin
Are we not all one under God
Why can't we treat all equal as God does
Can we not all be one as He created us
We all are brothers
All one
All equal
Under God's eyes
And under ours
All one
All one

March 26, 2004

Our Tongue

Oh, what lies within our mouth
Oh, that tongue
Out to deceive us
Our own tongue
Part of our body
Our very own body
An enemy to our very being
Or a mender of our health
Oh, what words form at the end of our tongue
Bringing forth health to our being
Or bringing forth sickness, death
Oh, what words flow from our mouth
Oh, how we speak out our very health, wealth
Oh, what words can do
Such powerful words
Oh, how we, who are made in the image of our Creator
Made in His image
Oh, the power to speak forth and bring about
Oh, us, made in the image of our Creator
Oh, in His image
Oh, what words He spoke forth
Became life itself
Oh, what words we speak
What do we speak out and bring forth
Oh, we must watch our every word
Our every word
Oh, how we can be deceived
To speak forth our own way of existing
We speak forth words that it was in my family
So therefore I have it
Or will have it
Oh, when do we speak forth the positive
Oh, watch our very words that flow
Oh, how we are bound by the tongue

Bound up—blinded we are
To not see what God had intended for us
Oh, how He gave Adam dominion
Dominion He gave him
We, now have the same dominion
The same authority
Oh, speak forth that which you desire
Oh speak it forth
It shall come to pass
Oh, speak it out
Speak it out
For my Lord sits at my Lord's right
He will bring all things to pass
It shall come to pass
Oh, use your tongue to speak forth the desires of your heart
Oh ye, made in the image of God Himself
Oh, where is your faith
Ye, one to whom I speak
Where is your faith

May 1, 2004

Our Tongue Deceives

Oh, how our tongue deceives us
What words we speak
Oh, how they hinder
Or
Help us
Oh, how we can say
I feel good
I am great
I am happy
I am satisfied
I am rich
And what we speak, these words are powerful
Us—made in the image of God Himself
Able to speak and bring forth what we speak
If we have faith in what we speak
And what then of the words we use
That are negative
I am sick
I am miserable
I am poor
I am unhappy
Oh, truly believing that to be
We speak that forth
Being tricked by the evil one
To keep us ever is his grasp
To keep us ever sick, poor, miserable, unhappy
For that is what we spoke forth
That is what we got
Oh, how easy it is to spell out the words of the negative
Oh, how hard it is to spell forth the positive
Oh, guard your mouth
Guard your tongue
For you are where you are on the ladder of life
Going up or down the rungs

Mostly because of your tongue, which direction you are going
Oh, watch what your tongue speaks
Oh, guard your tongue
For it is out to deceive you
To keep you bound in your sickness
Your unhappiness
Your poverty
Oh, even in your prayers, you speak forth the negative
Oh, change your thinking
Change your speech
Better off to lose your tongue
Than to be bound
Better off, born without a tongue
Than bound by what we speak
Better off to lose it I say

March 25, 2004

Our Walk

Oh, where is our walk when we walk with the Lord
Do we walk with Him and talk with Him throughout the day
Do we walk with Him in the cool of the night
What is our conversation
Is it of asking for this
Asking for that
Telling Him of hard it is down here daily
Or
Do we just walk and talk
Friend to friend
Lover to lover
Father to child
Hand in hand
Or do we walk and talk with our requests and complaints
Do you not know that He knows of our requests before we ask
Can we then not just spend some time with Him without the requests
Oh, can't you just walk with Me
Talk with Me
Yes, friend to friend
Lover to lover
Father to child
Can you not just walk with Me
I—who cares so much for you
I—who loves you beyond imagination
I—who watches over you
Protects you
Keeps you from falling
Constantly I have My eye upon you
Are you not the apple of My eye
Can you not take a moment of your time to walk with Me
Just a moment I ask

March 11, 2004

Our Worship

Oh, how do we worship our Lord
How do we worship the Lord of Lords
Oh, we have taken our Lord
Made Him a buddy
Someone on our own level
Oh, did not the prophets past
Lay prostrate before their God
Did they not fall upon their knees
Bow down before their Lord
And remove their very shoes, for they stood on Holy ground
Oh, where has our respect
To our Lord God
To the King of Kings
Oh, we come into His presence
We come into His presence
Too tired to stand before the King of Kings
Too taken up to raise our very hands before Him
Sit and yawn in His presence
Sit and yawn in His very presence
Walk in and sit down
We don't even have the decency to take our hats off before Him
Before the King of Kings
Before the God of the universe
The Alpha and the Omega
The Beginning and the End
We trail into His presence
Checking our watch
To make sure we don't spend too much time with Him
Make sure we are out in time
For our own agendas
Oh, here the King of Kings
We are coming to see
Coming into His house
Into His very presence

He is there, where two or more are gathered in His name
He is there
And where is our honor of Him
Where is our very worship
Oh, He is the Lord
He is the God Almighty
He is our Lord and Savior
And do we truly come in and worship Him
Do we come into His presence
And holy reverence who He is
Who He is
Oh, don't place Him at our level
A buddy so to speak
For he is our Lord
He is our King
And worthy of our praises
To place Him in His rightful position
To place Him high above all
To place Him above all
For He is worthy of His place
He is worthy of His place
Oh, give Him His rightful place
A place of true worship
Oh, we should worship Him
Oh, we should worship Him
For He is worthy of our praises
Our worship
And worthy for who He is
Oh, don't lower Him
Down to our level
Oh, lift Him high
Lift Him high
Oh, He is not at our level
He is not at our level
For He is high and lifted up
He is high and lifted up
Oh, worship Him with knee bowed
With your knee bowed

For He is high and lifted up
And worthy of our worship
Worthy of our praises
Oh, He is worthy
To stand in His presence
To kneel before Him
To bow down
To lie prostrate before Him
For He is Lord of Lords
He is the God Almighty
And One not to be reckoned with lightly

May 17, 2004

Out of Your Abundance

Oh Father, out of Your abundance
You bring forth Your gifts
Oh Father, your gifts
Presented to us
Your children
Out of Your abundance
Father, You give us gifts
Gifts to those that seek them
Oh, to those that seek them
Oh, Father, how You present Yourself to us
Coming in Your power and might
Bringing us into a genuine closeness with you
Oh, how You draw us in to Yourself
Pulling us closer and closer
Until we are one with you
One with You, oh Lord
As You are one with my Lord
So we too are one with you
As man and wife become one flesh
So we, Lord, are one with you
In completeness with You
Your total, Your total
Are we with You
Oh, how You pour out of Yourself
To those who seek you
How You give unto us Yourself
So that we become one with You
Oh Lord, my Lord
How great You are
To those that seek You and find You
For You are not far from us
You never were
You were always there
Always wooing us

Oh, when we finally stop and see You
For who You are
Oh glorious day for us
Oh glorious day for us, my Lord

May 1, 2004

Outer Courts and the Trinity

First Heaven
Outer court
Earth
On the outside
Waiting upon the Lord
Holy Spirit moves upon the people
Where cleansing is done
Where we prepare ourselves for the inner courts

Second Heaven
Inner courts
Upper Room
Spiritual realms
Battle ground
Where all await in expectancy of what is to come about
Where our Lord brings us through into the third heaven

Third Heaven
Enter into the Holy of Holies
Inside the secret place of The Most High
Only the purest
Only the washed may enter here
Where we meet with God
Face to face

April 3, 2004

Play Games

Oh, don't play games with Me
Don't play games
Come to Me with sincerity
With truth
Don' t play games
I will not stand for your games
Oh, stop your game playing
Stop it now
Come to Me in truth

March 14, 2004

Poise Yourself

Poise yourself ye people of the earth
Poise yourself
For I am sending out my messengers
They will carry My voice
They are throughout the earth
Oh, hear ye who has an ear
Hear ye I say

February 17, 2004

Praise Him

Oh, praise Him in the morning
Praise Him at the midday sun
Praise Him as the sun goes down
Oh, praise Him through the night
Oh, Praise Him
Praise Him
Praise Him
Oh Lord, You are every where are You not
You surround me
You are the very breath I breathe
The wonder of You
How secretive You are
Can one reach out and touch You
You are so hidden
Yet You circle about the earth
Your might and power is felt within the elements
Your presence is within Your creation is it not
Your breath is our breath
Oh, the wonder of it all
How precious You are
Oh, Praise Your Holy name
For You alone are Holy
Holy
Holy
Oh, Lord, does not the very earth cry out to You
And what of those who lie in their graves—awaiting—asleep
Oh, and what of them
Do they not cry out within their sleep
And what of those who ask for mercy Lord
And what of those
Oh, Father, I ask for mercy for them
Mercy
Oh, Father—incline Your ear to hear their cries
And what of those buried within the seas

Do they not cry out mercy from their watery graves
Oh, Father, who can know Your mind
Who can know the extent of Your mercy
Oh, I ask, Father for mercy for those that lie within their graves
Mercy, Father
Mercy

March 7, 2004

Pride

Oh, how pride seeks us out
Looking for a boastful person
Looking for the one that has placed himself high
Awaiting for the opportunity
To enter into such
Oh, how pride slides in
Elevating us
Oh, pride
Was it not pride
That took down the evil one
Was it not pride that entered him
And was it not this seed passed down
Through the generations
Pride
Oh, pride
You are a spirit
Are you not
An evil spirit
Passed down
From the evil one
Oh, don't allow pride to step in
For you too will fall
As did the evil one
You too will hit the bottom
You proud one

October 19, 2004

Prison

Oh, ye who fall into sin
Ye who are weakened and fall
Oh, ye who are bound in prison
Oh, how the evil one desires to trap you
Imprison you
Bind you in
Oh, how he keeps you trapped
How you settle into your prison
Settle right in
Accepting your position
Not even attempting to escape
Not even looking for a way out
Oh, ye who are trapped
Why do you remain there
Sitting within your cell
All bound up
Oh, look up, ye who are trapped
Look up
The door is open
The door is open
You are so blinded that you do not even see
Oh, the door is open
A way out
Get up
Awaken yourself
Stop sitting within your cell
Pitying yourself
Get up
Get out
You are cleared of any offence
Cleared, I say
You are bailed out
The door is open
Can you not just look up

See the door
See it has opened
Allowing you to escape
Why do you remain
Trapped
Who is telling you that you must stay
Who is whispering that you still must remain
That your offences are too great
That it is impossible that someone bailed you out
Who is telling you that
Oh, open your eyes
Open them
See the door
It is open
Open for you
Go, your offences are dismissed
Go out
Go out

April 7, 2004

Prodical Son

Oh, how the prodical son
After going his own way
When he came home
How his father welcomed him
How he went running to meet with his son
His wayward son
How he welcomed him
Bringing him into the house
Ordered the best of dinner to be prepared
How, he had the bath prepared
The clean robes to be placed upon him
Welcomed him with his blessings
His son
His wayward son
Has come home
Sat him down at the family table
Had him eat and drink with him
Oh, how he placed a robe around him
Put his arms around that wayward son
Oh Father, You our Father
Oh, when the wayward son comes home
To Your home
Oh Father, how You welcome him in
All is forgiven
How You put Your arms around him
Oh Father, they come home
They come home to You, Father
To Your house
Oh, Your forgiveness
Of that wayward one
You welcome them
Oh, how You have them cleansed
Place a clean white robe around them
They sit at the table with You

At Your table, Lord
They sit with You, Father
At Your table, Lord
Cleansed, wearing a white robe You have placed upon them
Cleansed, forgiven of their waywardness
Oh Father, how You have planned all
Your desire, Your will
That we all come home
That the lost
The wayward, stubborn one
Will come home
To Your house
Oh, how You are waiting at the gates
Your eye constantly at the gate
Looking afar off
You see this wayward one
On their way
Far away, but on their way
Oh, how You are waiting
Till they get closer
And Father, You go running to meet them
To meet them
And put Your arms around them
Pull them into Your arms
Tell them that they are forgiven
That You will not remember any of their past
But now they are home
Home with You, Father
Home within the family
Home with You, Father

May 12, 2004

Pull Nations

Oh, I will pull nation against nation
Country against country
Home against home
Father against son
Mother against daughter
Oh, I will bring this about
For one will turn against another
A people against a people
Oh, this will come about
Oh, be fearful of what is to come
Oh, be fearful of what is to come
Oh, hatred rides in the winds
Distrust follows its footsteps
Oh, what is about to come upon the earth
Oh, pleasantries about to leave
Trust about to leave
Honesty about to leave
Oh, what will come about the earth
Oh, hate, rebellion about to appear, not like ever before
Oh, what is about to come
Oh, there is a day about to appear
Not like ever before
Where the children of the evil one will rebel
Where the children of God will hide within the circle of the Most high
Oh, where safety will cover them
Where nothing can seek them out
Oh, what is about to come upon the earth
Oh, what is about to come

March 23, 2004

Quietness

Oh Father, how quiet it is
Amongst the noise
How quiet it is
Oh Father, how I miss Your voice
Oh Father, have I done wrong before You
Have I obtained pride in what You have given me
Oh Father, just to hear from You again
How I miss that, what we had together
How I miss that
What a void there is within me
Oh Lord, have I done wrong
Have I walked in pride
Oh Father, what have I done
Oh, how quiet it is
How lonely I am amongst my friends
Oh Father, talk with me
Oh Father, you have said to "wait upon the Lord"
Oh Father, that I will do, but Lord
How long shall I wait
How long before I hear from You
Oh Father, how we talked
How I miss that
I miss the wakening in the night
Where it was just You and I
I miss the times during the day
When You came to me out of nowhere
Whatever I was doing, You came
Wherever I was, You came
And now Lord, You say to wait upon You
Oh Lord, I am willing to wait
But, oh Father, how long shall I wait
How long will You keep me on hold
Oh Father, how long
For I feel the loss

I feel the loneliness
I miss the talks
I miss the closeness
Oh Father
Don't keep me on hold
Father, continue to use me
Continue to walk and talk with me
Oh, how I miss that
Oh Father, although You have said to wait
Oh Father, how long must I wait
Oh Father, in days past, You came to them
Some waited for years for You to come again to them
Oh Father, I am in my latter life
Oh Father, don't wait till I am nearly in my grave
Don't wait till then to come again as You have
Oh Father, use me now, while I am able
Forgive me if I have wronged You in any way
Forgive me if I have walked at all with pride
Oh Lord, I have not meant to wrong you
Or have walked in pride, but judge me if I have
And forgive me for doing so
But Father, continue to use this vessel
Continue to talk with me
I ask

March 29, 2004

Rebellious

Oh, ye rebellious youth
Where does your anger derive from
What anger you hold
What has brought it upon you
Was it not moments ago, when your love was purity and innocence was about
you
What crossed the line
Was it you or your parent
What brought about your rebellion
What brought about the barrier
Who made the first move
Was it of your parent whose love has never changed
They were still protecting, caring, loving
Or was it you, whose love has changed
Oh, you rebellious one
Your love has changed, you are stretching your power
Trying out adulthood before your time
A power of your learning years
Surely your educational years expound your parents
Does that give you reason to lower them in your and anyone else's eyes
Where has your respect gone, you rebellious one
Was it ever there
Or did it just up and leave
What has changed
What has changed
Was it your parent
Or was it you
Which has made the change in direction
You
Or
Them

March 10, 2004

Relax

Oh, how we like to settle in upon a chair
Hearing of pleasantries
Hearing of how wonderful everything is
Hearing of only nice things
Oh, how we sit back
Arms behind our head
Feet stretched out
While we hear of how wonderful we are
How beautiful we are
Oh, how we enjoy that
Oh, to hear that
Oh, how it perks our very heart
But
What if you hear the not so nice
Hear of your faults
Of your bitterness
Of your hatred for your neighbor
Of your infidelity
Oh, how that pricks our ears
How we then sit up and take notice
Defiance sets itself in
Oh, how anger comes upon us
Oh, we do not like to hear of anything that would upset us
Cause a ripple in our lives
Oh, how we hate to hear anything that does not pleasure us
Oh, how we just want falsehoods told
If it is the truth
Or how when we hear truth
We want falsehoods
Oh, how we are
We who settle into our chairs
Oh, truth will set you free
False will bind you up
Oh, do you wish falsehoods that are truths

Or truths and not falsehoods
Oh, open your ears ye relaxed one
Open your ears

March 25, 2004

Respect

Oh, what respect do you hold for the one that birthed you
To the one that nurtured
Taught
Cared
Loved
Protected you
Where would you be without her love
What shoes would you stand in
Does she sit within those you claim the highest of respect
Or have you tucked her away
Left sitting back
Waiting for a glimpse in her direction
Waiting a return of her love
Oh, where have you placed her on your list
Is she near to the top
Near the bottom
Half way
Or omitted completely
Oh, respect that women that gave you birth
Show your love
Help her in her helplessness
For one day you will walk the same walk
Be careful, young one.

March 6, 2004

Rush

Oh, how we rush about our daily lives
Busy, busy, busy
Taking no notice of the happenings of the day
Oh, the happenings of the day
Oh, blockages here
Blockages there
Oh, to find someone sincere
To find someone truthful
To find someone pure
Oh, where can they be found
Violence rattles the city
Robbery
Murder
Oh, violence about the city
Oh, we take no notice of the days
What signs are ever showing
Signs of the times
Signs to know the time is short
Oh, signs about us continually
Yet, we are too busy to read the times
Too busy to care
Oh, we must wake up
We must know the times in which we live
Oh, we must read the times
For time is short
We are racing around
Too busy to see
Yet, time is passing by
Time in which you must be getting prepared
Preparing your family
Preparing your friends
Your neighbors
Your city
Oh, we are in times when time is short

Time is very short
Or, you must check your list
Of things to do
You must check your list
For times are coming
Oh, times are coming
When there will be no time
There will be no time
Oh, see about you
See their faces
Faces of your family
Your friends
You neighbors
Your city
Oh, look about you
Take a look at their faces
Faces of people about you
Who do not know of Me
Who do not know of Me
They must be told
They must be told
They must be told
Oh, see their faces
Oh, you must go out
You must tell them of Me
You must go and tell them
Oh, time is short
Times is so short
To save the lost
To save the lost
Oh, see their faces
See their faces
For I see their faces continually
I see their tears
I feel their cries
Oh, you must go and tell them of Me
You must

May 8, 2004

Saints

Oh, for the prayers of the saints
Do they not bring the children home
Do they not find the wanderers
The lost
And bring them home
Oh, for the prayers of the saints

March 12, 2004

Saved

Oh, Lord, You have said
If we confess with our mouth
And believe in our heart
That Jesus Christ is Lord
That we are saved
Oh, Lord, You have said that
Oh, Lord
Will not every knee bow
Every tongue confess
That You are Lord
Oh, Lord, if then every
Does that mean every
Every I say
Does that mean that all that confess You to be Lord are saved
All I say
All that confess
All that bow and believe in their heart that Jesus My Son is Lord
They will be saved
All
And all them that confess, Lord, then are saved
They are saved

March 28, 2004

Sheep in the Pasture

Oh Father, are not we Your sheep
Your sheep within Your pasture
But Father—the gate is open
To allow Your sheep to go in and out
But Father, the gate is open
Allowing the evil one to come into the pasture too
Allowing him to mingle amongst the sheep
Mingle with them, Father
Eat with them, Father
Oh, how he swirls through the flock
Looking and checking for the weakest lamb
For the sickly ones
So that he may snatch it from the flock
Oh, how he looks just like us
Oh Father, how he wanders through the pasture
Looking at all Your sheep
Oh, the gate is open, Father
The gate is open
Oh, beware, you weakest of sheep
Oh, beware, you that are sickly and weak
For the evil one is wandering through the camp
Wandering in and out of the camp
Looking for what he desires
Looking for what he desires
Eating with us, he is
Looking to snatch away
But Father, You the owner of the sheep
Watching over Your sheep
You are ever there with Your staff in hand
Never is Your eye closed
You are always watching
Oh, yes, the owner sees the evil one
Oh, He knows he is roaming amongst the sheep
Yes, He knows the gate is open

Open to bring in the sheep from distant lands
But, He sees too, any of the wanderers that wander out the open gate
He searches for them
Calls to them wherever they may be
They hear Him calling
Calling them back
Calling them in
Brings them back
Ever protective He is of His sheep
Ever watchful of His sheep
Oh, the evil one may wander in and out
But not one lamb will be snatched from the fold
Not one lamb be led away
Not one lamb devoured
For all the sheep are numbered
All wear the stamp of the owner
All are there
And though the evil one roams in and out of the camp
Though he eats with the lambs
Not one will be lost
No not one
For all belong to their owner
All have the stamp of the owner
Stamped from the beginning of time
Stamped I say
All that are stamped belong to the Lord
All that are stamped
The gate will close when all the sheep have come into the camp
The gate will close

April 1, 2004

Signs of the Times

Oh, ye can read the signs of the times
You know the time is short
Yet, you continue to go about your life
Treating each day the same
Yet, you can sense a charge in the days
Life is not
As you once knew it
Oh, anger fills the people
Oh, selfishness fills the people
Oh, where is the discipline
Oh, how children flare back at their parents
Respect has fallen away
Oh, children unruly
Parents doing their own thing
Oh, rushing here
Rushing there
Oh, no time for the sick
No time for the elderly
Oh, no time for each other
You go, go, go
Oh, you see what is going on about you
You see the times
You know time is short
Yet, is there any change in you
Is there any change
Oh, no different from yesterday
Everyone runs around, thinking of self
No concern for those about you
No concern for those you love
No time for any of them
Oh, you buy your children's love
Buy them this, buy them that
The latest of attire
The latest of play toys

But soon they reach adolescence
And you have lost them
They no longer are the little child
They no longer listen to you
They rule themselves
They rule you
Oh, they are a rebellious generation
Oh, where is the discipline
Oh, what has happened to the young children
Where has their innocence gone
Oh, we parents
We, are to blame for their behavior
We are to blame
We did not use the rod
We did not use the rod
Oh, where are the children now
Oh, unruly ones
Unruly ones
Oh, too late to turn back time
Too late to use the rod
Oh, is it not written that the children will turn against their parents
And parents against their child
Oh, you see the signs of the times
Oh, prepare for days ahead
For your unruly children will have no thought
To cast you aside
No second thought what they will do with you
Oh, if only time could be turned back
Too late for that
For the mold has been set
The mold has been set

May 6, 2004

Sit in Wait

Oh, Lord, I sit in wait
In wait in Your upper room
In wait for Your promise
Your promise of Your power
I sit in wait, My Lord
Sit in wait
Oh, what You have in store for me
Oh, what You have in store for me
I shall be ready
With my lamps trimmed
I shall be ready
Oh, Lord, I will be ready
For when You come in Your power and might
With Your gifts
With Your gifts
Oh, I sit in wait
Oh, as your disciples sat in wait
For Your promise to them
I too sit in the upper room
Awaiting Your promise as well
For when Your promise came
The room was filled
Filled with Your glory
Filled with Your Spirit
Oh, what power was then endued
What power they received
Oh Father, I too sit in wait
For Your promise to me
I sit in wait
My mind on You
My heart on You
Oh, Father, I await
I await in Your upper room

I will not leave it until You come with Your promise
I will not leave, My Lord

April 8, 2004

Sit, Be Quiet

Sit, be quiet, listen
Listen to what I have said
Sit in the quiet place
Open your heart, your mind, your eyes
Have I not told you what I will do
Have I not warned of days to come
Take heed, the day is coming
The days I spoke of are soon to appear
Look for my signs
Listen to what I say
I have told you of the days to come
There will be many that I will spew out of my mouth
Take heed, be always ready
For you know not the time
Be prepared—those days are just ahead
Don't sit back resting
There is no time to rest
Days to warn of
Days of fear and fright
For I am coming soon
Be ready

February 3, 2004

Snuffed Out

Oh, how quickly our life can be snuffed out
Breathing one second
Gone the next
No time to ask for forgiveness
No time to cry for help
Snuff—gone
No chance to come back
Start over
Try again
No second try
Snuff—gone
Oh, how quickly we can be taken
How quickly
Perhaps we are a parent
Perhaps a child
Perhaps alone
When we are taken
Snuff—gone
Oh, where do we stand with the Lord
Are we ready to meet face to face with the One who gave us life
Are we ready
What will be written in our books
What all will be written for all to see
Oh, we step through the portals of time
From life
To death
To face to face with our Maker
Snuff—gone
Oh, where do you stand with your Lord
Oh, always keep your slate clean
Be prepared, be prepared
For your next breath may be your last
And then what do you face
Oh, prepare yourselves

Wipe your slate clean
Keep it clean
Snuff—gone

March 22, 2004

So High and Proud

Oh, ye who stands so high within My court
Oh, ye who stands so high
Oh, you have placed yourself there
Ye, who thinks yourself so high
Oh, where did your highness come from
Oh, ye who stands so proud
Oh, ye who are so proud
So high and mighty and proud
Oh, I will pull you down
I will drag you down
Down, down, down
Down to the lowest of them all
I will drag you down
Oh, I detest the high and mighty and proud
I detest
I detest
Oh, your proudness is not of Me
Your highness not of Me
Oh, ye who placed themselves so high
Oh, so high you sit
So high you sit
Oh, you take the highest chair
You take the highest stand
Oh, I will drag you down
Down, I say
Your proudness I will wipe off your face
Your proudness I will remove
Oh, you will have nothing to be proud of
For you will be the lowest
The lowest
Dragging yourself along the ground
Oh, ye who stood so tall
I will smite you down
Down to the ground

Along the dirt you will crawl
Crawl along the dirt you will crawl
Dragging yourself along with the crawlers of the night
Along with the crawlers you will crawl
For your proudness brought you down
Your highness brought you down
I will smear your proudness from your face
Oh, ye will be the lowest of all
The lowest
Oh, ye who thought themselves so high
So high
Higher than the Highest
Mightier than the mightiest
Proud ye were
To stand in the highest place
To sit in the highest seat
To step up in the Holiest place
Oh, ye proud one
For I have yanked you down
Oh, I have yanked you down
I have yanked you down
Now you crawl along the earth
Dragging yourself
Oh, where is your proudness now
Oh proud one
Oh, where is it
Oh, your proudness now gone
Your place of power now gone
Now, you no higher than the earth worm
No higher
All is left of you is your will
Your evil will
To drag down any with you
To pull them down, along the ground
To pull them to your lowest level
Oh, where did your evil come from
Ye evil one
Where did it come from

Oh, I will destroy the evil from you
I will destroy it
I will destroy the high and proud
I will destroy all evil
It will remain no more
No evil will reside with Me
No evil
No place for the proud to stand
No place for them to sit
For no evil will be found with in My kingdom
No evil
For evil will be destroyed
Destroyed I say
Destroyed

April 10, 2004

Soaring

Oh Lord, You soar within the winds
You fly like the eagle above the winds
Floating above the clouds
Down, through the clouds and up again
Up unto the highest mountain
You cast Your eye upon the earth from up there
And off You go again, up and up into the skies
Down through the clouds and up again
Oh Lord, how You move, how You soar like the eagle
Your eye sees all
Oh Lord, You are everywhere
See the tiniest speck from up there
You can fly down and catch it before it reaches the ground
Oh Lord, how great You are

March 19, 2004

Species

Are we the only species God created that sickness and diseases reign
What other species did He create where sickness and diseases envelop them
Oh, the other species have sicknesses and diseases
But not as we do
Oh, evil one—you, the master of sickness
The master of diseases
You desire nothing of God's other species
There is nothing found in them that you desire
You slither your way across the lands
Across the skies
Across the waters
Slithering your way with sickness and diseases trailing your backside
You, out there looking for where to plant your wickedness
There is no desire for you to burden and downgrade the animal kingdom
Nothing in it for you
They can't run and do your beck and call
Nor can the birds of the air
So you do your slithering, finding one that perhaps is weakened, or one you
wish to pull down yet further
You bring upon that one, any sickness you can throw at them
But oh, ye evil one
You do your best, but our God turns it around
We will still praise our Lord
We will still lift our hands with joy
For sickness and disease may encompass me
But I know my Lord reigns
He will bring about good through it
Oh, evil one—slither away from God's children
For there is no prize to be won here
No prize at all

March 21, 2004

Spirit of Sickness

Oh, you spirit of sickness
What name do you run with
And what of you, oh spirit of disease
What name do you run with
How you slide yourself in
How slick you are
How do you motivate your entrance
What schemes do you use
Oh, where do you get your orders to act
How you settle yourself in
Take placement within one's body
And what name do you run with, oh spirit of sickness
What name
When did you take up residence
Did you come in suddenly
Or did you come in with the generation been passed down from one to
another
Oh, where do you go when your vessel turns to dust
Where do you go
Do you lie in wait as the next available vessel passes on by
Oh, how you move in
Bring with you all you can
Take over the vessel do you not
You spread out yourself encasing your vessel with your tendrils, encasing the
whole being
Oh, what name do you run with, oh spirit
What name do you have
Where do you make your home
Is it in one's brain
Organ
Blood
Flesh
Bone

Oh, where do you hide
Who allows you to take up residence
Who

March 8, 2004

Spiritual Gifts

Where can one go to obtain spiritual gifts
Are they purchased
Are they borrowed
Or
Are they given
And what do I do when given a gift
Do I tuck it away and hide it
Am I ashamed of it
Do I abuse it
Can I return it for another
Oh, what do I do
If I don't use it
Is it given to another
If I am ashamed does it leave
Oh, what does one do with their gifts

March 10, 2004

Spiritual Realm

Oh, to see inside the spiritual realm
Oh, to be able to see
How do they move around
Do they fly
How do they travel
Do they "hitch" a ride
Or does the wind move them
Oh, do they sit upon chairs like us
And do they lie upon a bed
Oh, how do they move
What pushes them
Do they just float awaiting your move
Or are they sent directly to you
Where do they get their orders
What if they cling to you
Can they crawl inside unknowingly
Then what do they do
Do they laugh with us
Or at us
Do they disturb us
Or do we disturb them
When we are ill—are they ill too
Oh, what type of spirits are there
Are there good ones
Are there evil ones
Oh, what kind attached itself and crawled into you
What kind
Oh, to be able to see into the spiritual realm

March 9, 2004

Storage

Oh, prepare yourself for days ahead
Prepare ye yourselves
For days ahead of famine
Oh, you have stored up
Stored up
Days ahead will come quickly
Oh, you placed your seed within your storehouse
Oh, placed it there
Oh, you stored it up
For days ahead
Those days will come quickly
Oh, your stored-up seed
Placed it there for safekeeping
Oh, don't let the robber steal it away
Oh, he will try to steal your seed
Oh, you have stored it up
Safely you stored it
Oh, for what use is it, stored in the store room
Where it can't be used
Nor will it grow
Oh, how dried it is in the store house
Oh, what a valuable gift your seed
Oh, safely stored
How will it grow
Oh, what a vast amount of seed is missed
Oh, what your seed could have produced
Oh, you safely stored your seed away
Kept it not only from the robber
But kept it from the hungry
The hungry
Oh, how you let many die without food
Oh, they die continually without food
Oh, what days lie ahead
Days when your seed will be needed

Oh, the mouths your seed will fill
Oh, how full they will be with your seed
Oh, take it out of the storehouse
Where it will grow
Produce more seed
Enough seed to fill the hungry
Enough seed to keep them from death
Oh, such famine lies ahead
People dying for lack
Dying for lack
Oh, bring out your seed
Feed My people
Oh, there are more seeds within the storage room
Seeds that could have multiplied
But hidden they were
Hidden and not allowed to grow
To produce
Oh, we will need all the seed for days ahead
Oh, hungry people
Hungry people
Oh, feed them, My children
For you have an abundance
Oh, when famine comes
Oh, have your seed out of the storage room
Seed that has produced
Enough seed to feed the hungry
Oh, bring out all your seed
For seven years will be famine years
Famine throughout the lands
Oh, bring out all the seeds now
Feed the people
Oh, feed the hungry
Oh, fill them
For they are dying for lack
Dying for lack
Dying for lack
Oh, days ahead many will die for lack
Many will die for lack

Oh, bring out your seed now
So it will multiply
Multiply to feed the people during the famine years
Oh, bring it out now
Bring it out now

April 21, 2004

Streets of Gold

Oh, to dance upon the streets of gold
To dance upon the streets
Oh, to dance before the Lord of Lord
To dance before the King
Oh, to dance upon the streets of gold
To dance before my King
Oh, bring your timbrels
Oh, bracelets upon your feet
Oh, make a joyful noise unto the Lord
A joyful noise to Him
Oh, dance upon the streets of gold
Oh, dance before the Lord
Oh, bow down before the King of Kings
Bow down to the Lamb of God
Oh, sing before the Lord of Lords
Sing before the Lord
Oh, come before the King of Kings
Oh, come before Him now
For he is the Lord of Lords
He is the King of Kings
He is he One who washed you clean
He is the One who presented you before the King
Oh, sing before Him, all ye who stand
Sing before the Lord
Oh, dance before the One of One
Oh, see who He really is
Oh, come before Him all ye people
Come before the King
Oh, look who has brought you here
Oh, look upon His face
Oh, look, ye people of the earth
Look upon His face
He is The One
The only One

That has saved you by His grace
Oh, look upon this face ye one
Look upon His face
For He alone has brought you home
By His saving grace
Oh, look upon the lamb of God
Oh, look upon His face
Oh, see the Lamb who brought you home
And saved your very life
Oh, see His face, His Holy face
Who died to bring you home
He gave His life
His earthly life
His desires like your own
He set them all aside
His feeling, His hurts
His loved ones
His earthly all
He set them aside
Aside
He stepped aside, so you may live
He set them all aside
For you, My child
He set them all aside
So you may live
Oh, see what He has done
Oh, see how He stepped down from His throne
Took His place amongst man
He walked our walk
Knows us as we are
Knows our weaknesses
Our strengths
Oh, how He knows
Yet He remained sinless
To bring us sinful people home
He held back His feelings
His thoughts
His all
To bring us all home

Oh, see His life
His life given for you
He breathed as you breathed
He felt as you feel
He had parents as you so do
He had feelings as you so do
Yet, He put them aside, for you
He put them aside for you
Yet you curse His very name
You laugh at His name
Oh, be ever watchful
Ever watchful
For this is My son
My only Son
My very Son
Who you laugh at
Who you curse at
Oh, beware, ye who do this
Beware, for this is My son
My only Son
I am very protective over Him
Ever watchful over him
Oh, beware, ye who laugh at Him
Curse His very name
Oh, beware, ye who do this
Oh, how wonderful You are
Oh, can anyone stand up to You
Oh Lord
Can anyone step up to the plate
And declare he can do as You have done
Oh, Lord
Forgive us here on earth
Forgive us who fall into sin
Oh, forgive us for we are born into sin
Oh, Lord
Re-mold us
Re-make us
To be as You, oh Lord
For we are Your children

Born now into Your family
Oh, Lord
See our weaknesses
See our faults
Bring us home
Oh, Lord
Bring us home
Washed by Your blood
Cleansed
Cleansed to be able to walk before the Lord
Walk into His chambers
Presented pure before Him
Pure before the Lord
Oh, how wonderful You are
To see before the foundation of the earth
That we were a chosen people
A chosen people
A circumcised people unto the Lord
Brought into a covenant with the Lord
Brought into a covenant with the Lord
Children of Abraham
All children of Abraham
All bought, now numbered within the sands
All numbered within the sands
All are under His covenant
All are His
Oh, praise His Holy name
Praise His Holy name
For we are bought
We are His
All children of the Lord
All are His
All bought
All His
Oh, praise His Holy name
Praise His Holy name

April 7, 2004

Take for Granted

Oh, how we take things for granted
Oh, we take our health for granted
We take the health of our loved ones for granted
We take our safety for granted
The safety of our loved ones
Oh, how we even take our salvation for granted
Oh, not even thinking perhaps of the many prayers of our loved ones
Oh, the prayers of our ancestors
Oh, what prayerful people they were
We take so much for granted
Not even thinking perhaps that a loved one
A friend
An unknown
Cried unto the Lord for our salvation
Oh , their prayers now answered
Perhaps our prayers now answered
Oh, how often do we stop and thank our Lord
For His mercy
His love
His love that none should perish
Oh, how He tingled the ear of a loved one
A friend
An unknown
To pray for us
Us, a lost soul
To bring us to a point of recognizing our sin
And following the prompting of the Holy Spirit
Oh, we too must pray for family
For friends
For the unknown
For they are out there
Needing our prayers to go forth
Needing our prayer to step into the gap
Oh, we too must recognize the call of the unsaved

For we were there at one time
And perhaps if not for the prayers of a child of God for us
Where would we be
Oh, be sensitive to this, children
Be sensitive for the unsaved
For they are all about you
Oh, when you walk down the street
Take a look at the people passing
How many are a born again child
How many are lost
Oh, a cry goes forth for the lost
A cry goes forth for the weak
Oh, a cry goes forth for they that have wandered too far from the fold
Oh, children—sharpen your tools
Oh, the catch is ready
Oh, cast your net
Oh, your net will be full
Oh, cast it forth unto a lost people
Oh, see them
For they too must come into the fold
They too must be directed to the path
Oh, they must come in
Oh, I wish for none to perish
None to perish
None to perish
Oh children, you are My vessels
You are My vessels
Oh, care for your brothers and sisters
Oh, you must care for them
For they too are My children
A lost generation
Oh, you must work hard
Sharpen your tools
Bring in the lost sheep
For joyous is the day when they all come in
Oh, joyous day
Oh, the children of Abraham
Counted as the sands of the sea

Oh, where is the number
Oh, all My children
All are coming into the fold
Oh, bring them in
Bring them all in

April 28, 2004

Temptations

Oh, how it plants itself around us
In front of us
Beside us
Breathing down our neck
Whispering continually in our ear
Look at this
Look at that
Taste this
Taste that
Buy this
Buy that
Oh, how we fall into temptation
Looking
Smelling
Feeling
Tasting
Hearing
Oh, how our five senses tingle
When approached by temptation
Oh, how weak we are
Falling into temptation
Not really needing this or that
Not really wanting to taste this or that
Not really wanting to feel this or that
Not really wanting to hear this or that
But, it whispers into our ear
Oh, how weak we are
To listen and then fall into temptation
Oh, it is out there, everywhere
Oh, resist temptation
For your satisfying appetite will not stop at just the
One look
One feel
One taste

One smell
One listening
Oh, we want more, and more
This is what temptation does
Once it has planted itself into you
And you fall into her
You will continue with more temptations
Oh, get bold with yourself
Refuse, resist temptation
Start off the way you started, start with one
Then two and on.
But start now, for you will drown in your temptations
Get out now
Get out now

March 26, 2004

The Birds

Oh, the birds of the air know that winter draws nigh
And prepare themselves to fly to warmer weather
Yet, you see that the season is about to change and don't prepare for it
Oh, the birds of the air have better sense than you

March 13, 2004

The Blind to See

Oh Father
How You are allowing the blind to see
The deaf to hear
The lame to walk
Oh Father
You—Father
Opening the ears of the deaf
So that they may hear Your voice
Opening the eyes of the blind
So they may see
Oh Father
How the lame will walk
Oh, when the blind now read My word again
They may see
Oh, when the deaf hear My voice again
They will hear with clarity
How the lame walked
Now, they will walk with boldness and authority
Oh, how the blind now see
The deaf now hear
The lame now walk
Oh Father
You are doing a mighty work in this day
Oh Father
A mighty work You are doing
A mighty work

March 28, 2004

The Clock is Ticking

The clock is ticking
Time is running out
The minutes slip idly away
There will be no time left
I will not hold back from what I intend to do
No changing of My mind
I will come suddenly
Suddenly, I say
Your life will be plucked away
I will show you My might
You have your gods before you
I am your God
I will not share with nothing else
I am your God
My patience is coming to an end
You glace at your measly life
Who gave you that life
I will tell you
I will not hold back not much longer
And you will run
You will hide within the rocks
You cannot hide from Me
No darkness can I not pierce
I will seek you out
You think I cannot see
I see your every move
Your life is mine
Mine, Mine
Turn to me, now I say
Your minutes are slipping away
I am coming

February 4, 2004

Oh Evil One

Oh, does not the evil one walk in each step I take
My left, his left
My right, his right
Oh, how he follows, awaiting
Waiting for me to trip
Waiting for me to fall
Constantly he is on my back
Pushing me this way and that
Oh, how I stand firm
Firm I stand upon my feet
For He who keeps me
Keeps me from tripping
Holds me up lest I fall
Oh, how firm I stand
Oh evil one

March 16, 2004

The Heavens

The heavens are being stirred
Times are coming
A bringing in what has been planted
Harvest time
Walk over that which did not mature
Cast aside that which did not take root
Gather in the ripe plant
Leave that which did not take
Leave the plant for further nurturing
I will nurse the weakling
I will strengthen its roots
Then I will gather that to Myself
Beware of days ahead
There will come those that speak My word
Oh, test the spirits
Oh, test to see if they are mine
Oh, check My word
Dark days lie ahead
Days of deceivement
Prepare yourselves
Oh, know My word
Oh, test the spirits.

February 21, 2004

The Heavens

Oh, can one know the depth of the heavens
Does one know their height
Does one know how wide they stretch
Does not man know they are measureless
I have stretched out the heavens
I have placed earth as a mere speck within the heavens
A mere spot amongst the stars
Does man think he can reach into the heavens and look beyond the stars
Oh, a foolish people

March 3, 2004

The King

Oh, do you see the King
Do you see our Lord
Oh, riding in the skies
Oh, do you see Him coming
Oh, His Majesty
Oh, see Him
All ye on earth
For every eye shall see Him
Oh, every knee will bow
Oh, the greatness
The magnitude
Of those who ride with Him
Oh, see Him
Oh, every eye
Oh, every knee
Oh, all will bow
All will confess
That He is their Lord
All shall confess
All shall bow
All will say
That He is Lord
He is Lord
He is Lord
Oh, rising with Him
Oh, rising with Him
Oh, rising with Him
Oh, my Lord
My Lord
My Lord

March 25, 2004

The Majesty of My Lord

Oh, the majesty of my Lord
How vast Your universe
Can one see Your vastness
Your power
Your might
Oh, how utterly small we are with in Your great universe
Oh, Father when I see the hugeness of Your angels
Just how huge You must be
Where are You hidden behind the curtain
Can one pull back the flap and peek inside
Just a glimpse of Your glory
Can one stand after looking within the walls
Can one live having seeing the majesty of it all
Oh, how huge You are
Oh, to have a glimpse
Just a glimpse
Pull back the curtain
Allow me to peek
Just a sliver of an opening that I may see
Oh, how huge and mighty You are
How small we are within the heavens
Yes, I can see how You hold all things in place in one of Your hands
Oh, I can see

March 4, 2004

The Marriage

Oh, the word has been sent out
The word of the impending marriage
The marriage of the bride and Groom
Oh, the atmosphere is glistening with the aura
Oh, all creation is in a standstill
Oh, all feel the atmosphere
Oh, the electricity is everywhere
Oh, the announcement has been sent forth
Oh, the announcement has been sent forth
Oh, we, the bride, have picked up the notice
We have picked up the notice
To start the bathing
To start the cleansing
Oh, we must soak in the waters of perfume
Oh, we must be clean
Oh, the scent of the perfume is everywhere
The air is filled with the scent of it
Oh, how preparations are being done everywhere
Oh, a readiness is abound
Oh, have all the invitations been sent out
Is everyone on the list that we required
Oh, what of those that did not get an invite
Do they not even get a chance to stand in the alcove
To be there, even though not invited
Oh, what of those
Oh, make room for them all
For many wish to come to the wedding
Make room for them all
Oh, which dress is most appropriate for me
Which would be best to wear
Oh, the excitement of it all
I feel the tension
The laughter
The happiness

Everywhere
Oh, it is all around me
All seem to be caught up in the excitement
Oh, all the invitations
Have been sent out
Oh, make sure they all got one
Oh, make sure they are all coming
For room has been made for them all
Oh, room has been made for them all
Oh, a larger room has opened to accommodate them all
Oh, ask them all to come
Tell them not to worry about their dress
For they are invited to come
Tell them not to worry to bring a gift
For all is covered
All is covered
No gift is required
Oh, the dress is not a matter
Just come to the wedding
For a glorious wedding it will be
For it will be a wedding for all
Oh, the air is filled with the glory
Oh, I feel the excitement
I feel the urgency of it all
Oh, invite everyone
Don't leave anyone out
Oh, the preparations are done
All is done
Only the announcements to be read
Only the announcements to be read
All things are done
All things are accomplished
Oh, prepare for the wedding
Prepare for the feast
Oh, bathe yourself
Perfume the waters
Oh gown yourself
For the wedding is about to take place

The engagement period now over
The engagement period now over
The cleansing now of the bride
The dressing of the bride
Oh, all things in place
Oh, make sure the invitations are all out
For the wedding is about to take place
It is about to take place

May 3, 2004

The Mysteries

Oh, the mysteries of the universe
The mysteries of it all
What holds the stars in place
And are the sun and moon the same

Oh, the mysteries of the universe
The mysteries of it all
What holds the earth in place
And are the planets the same

Oh. the mysteries of the universe
The mysteries of it all
Where does the heavens end
And where do they begin

Oh, the mysteries of the universe
The mysteries of it all
Where does my Lord live
And where does He not

March 8, 2004

The Sounds

Oh, the sounds within the heavens above
Oh, the sound of thunder roaring
The lightning shooting its forks upon the earth
The hail that bounces off from whence they landed
Oh, what is going to come upon the earth
Oh, this is child's play to what is going to come upon the earth
Oh, the devastation that will come with the winds
The devastation that will come with the rains
Oh, what will come upon a people
A people that do not know their God
A people that rejected Him so
A people that held out too long
Oh, what will come upon a people
Oh, pity the people
A people that held out too long
Oh, pity them
Oh, pray for this people
A people that held out too long
Oh, send your prayers out for them
For Your prayers will bring them in
Oh, your prayers will bring them in
Your prayers will bring in a people
A stubborn people
Who rejected a calling
Rejected a time
Rejected their Lord
Oh, send out your prayers for this lost people
A lost people who now run and hide
Who run and hide beneath a cover
A cover that will hide them from their Lord
They think
Oh, like a turtle
They pull themselves into a cover
Hiding themselves from the outside

Oh, who can hide from the Lord
Who can run and hide from Him
Oh, a time is coming, when they will run
They will run and hide from the Lord
Oh, who can hide from Him
Who can hide from Him
Can He not see through all
Can He not see all
Oh, who can hide from Him
Does He not know everything
All
Oh, who can run from the Lord
Oh, come now to Him
Come now to him
Don't be forced to come to Him
Don't be forced to come to Him through prayers
Oh, come to Him on your own accord
Oh, He loves you so much
Wishes for none to perish
His wish for none to be lost
For none
Oh, His wish
His will
For none to be lost
Oh, I pray for those now stubborn
I pray for those that have rejected
I pray for them, my Lord
That they too will come to You
That they too will come with singing
And dancing
And praises
For You alone are worthy
You alone are worthy
Are worthy to come to
To come to, my Lord
For You alone are worthy for us to come

May 9, 2004

The Spirit

Oh, the Spirit of the Lord is moving upon the earth
Moving this way and that
Settling here
Settling there
The Spirit is moving
Oh, latch on as He passes by
Latch on
Hold on
Don't let go
For He is moving across the lands
Moving, never staying too long in one place if not welcome
Moving on to find the welcome mat placed at the front door
Oh, He is moving
Latch on
Hold on
Don't let go
Don't let go

March 14, 2004

The Trumpet

Oh, the shout of the trumpet is near
Oh, the shout of the trumpet is near
Oh, the shout of the trumpet is near
Oh, ye inhabitants of the earth
Oh, ye inhabitants of the earth
Oh, don't be found lukewarm
Oh, your hour of decision is now
Don't linger
Oh, there will be trials and tribulations like never before
Don't be caught standing in the middle
Don't delay
No time to be warned
Oh, times are coming like never before
Times of trials
Times of tribulation
Oh, like never before
Like never before
Oh, ye inhabitants of the earth
I wish for none to perish
None to be lost
Oh, ye inhabitants of earth
Make your decision now
Now, I say
For the wooing of My Spirit may not always be with you
Oh, make your decision now
Don't be lukewarm
Oh, come to the other side
Come now
Oh, time is running out
No second thoughts
I wish for none to perish
None to be lost
Oh, come while there is time
Come while there is time

For times are coming
Times like never before
Oh, come to Me now

April 16, 2004

The Wanderer

Oh, we are all wanderers
Wandering the desert
Circling back and forth
Oh, how long do we wander
Searching for the land of milk and honey
Oh, we wander and wander
Oh, some of us wander the forty years
Some of us wander less
And many take up the forty years plus
Oh, we search for that which our God promised
Oh, we search
Oh, we His children
But we wander
Looking for the promised land
Land of milk and honey
Oh, how He feeds us over the years
Cares for us
Goes before us
Guides the way by day
Guides the way by night
Watching our every move
But we grumble
Wander and wander
Over the sands we go
Trailing back and forth
Dragging all our baggage with us
We circle and circle
Up over mountains
Down into the valleys
Around and around we go
Grumbling constantly
Where is this promise our God has promised
So we remain the wanderers
Oh, when do we stop and realize we missed the promised land

So many times we passed it by
So many times over the years
Passed right by it we did
Oh, His promise
This land where we can rest under Him
Where we can call it home
Where His eye ever is
Oh, is it really a land
Oh, is it truly a land of milk and honey
Where we will live peacefully
Oh, is it really a land
Oh, many do not ever make it
To this land
Oh, until we fully come to Him
Fully and earnestly come to Him
Get rid of the old baggage
Get rid of our old gods
Get rid of our old ways
Will He open our eyes
Open our hearts
And guide us to this land
Oh, we have escaped our slavery
Oh, we have escaped our slavery
Oh, we enter into the land of milk and honey
God's promised land
Oh, we will still have our tears
Our troubles
But we have now crossed the river
Entered into His land
His covering
His love
His protection
For He has made a trail for us
He has made the sacrifice for us
He has made all things possible
He has done it all
He has opened our eyes
He has opened our hearts

Oh, the promised land was always there
Always there
Yet, we trailed on by so many times
Blinded by the sands
Oh, we now into His promise
His love
His all
Saved from slavery we are
Saved from that which bound us
Saved we are

May 8, 2004

The Winds

Do you know where the winds start and where they end
Only I know their beginning and their end
Only I
Who can say if they came from the east and moved to the west
or from the west and moved to the east
I move the winds
I know from whence they came
Only I
Oh, dance before your King, my people
Dance
I take pleasure in your praises, your dance
Oh, you dance before man
I am a jealous God
Dance before Me
For I see your hearts
I know your loves and your sincerity
Oh, cleanse yourselves, my people
Purify yourselves
Wash your house clean
Remove the debris
Dust out the corners where dirt will hide
Mop your house clean
Then, I will come
I will come with power and might
Oh, purify yourselves, My people

February 29, 2004

The Wind That Blows

What is in the wind that blows
What does it contain
What spiritual forces move and ride upon the wind
What spiritual force
And how about the words that are spoken throughout the earth
Are they not carried within the winds
Where do our words go after leaving our mouth
Do they just float around too
Do they just evaporate
Do they fall back onto the lands
Or
Do they go up into the Heaven of heavens
To the Throne of God Himself
Oh, what words have we spoken

March 10, 2004

Thoughts

Oh, where do your thoughts run to, My son
Do they run to Me
What evil lies yet dormant within your bowels
Oh, flush it out, My son
Is there any evil that man has not conceived
Oh, run, My son
Run unto Me

March 6, 2004

Time Is Short

Oh, the time is short
Time that was allotted is now stretched
Oh, that You give us time to go throughout the lands, telling them of Your love
Oh, can anyone hear
Can anyone grasp what is been said
The Word goes out
Does anyone hear
So discouraging
So discouraging
Look around you, are there ones round about you that need My message
Give them My message
Let them know that there is love
That I am here to help and deliver
Oh, let the people know that there will be times ahead that they will need My word, My power
Oh, go tell the people
Stand on the mountain tops, declare with a loud voice
The day of The Lord is about us
Declare the word, so that everyone may hear
Oh, shout it from the very tip of the mountain
Declare it from with in the skies
Oh, declare My word to every ear that will receive
Oh, shout it out, so that the nations will hear with one accord, that the day of the Lord is about them
Let them know
Time is about them
Let them know
Wipe your feet after you tell them of My love
My mercy
My power
Wipe your feet I say
Wipe the dust off
Do not return to that land

Return to Me
They have made their decision
But let them all know first
Let them know

March 4, 2004

Time Ticking

Oh, hear the sound of time ticking away
Each tick a moment leaving your life
Each tick—consider a beat of your heart
Tick—tick—tick
Time leaving
Never to be replaced
Time going by
Each tick, each heartbeat, each breath
A moment in your life
Oh, consider where you stand in the moment
This heartbeat, this breath
Tick—tick—tick
Oh, are all things right
Tick—tick—tick
Oh, where is your stand in this moment
Tick—tick—tick
Oh, time is moving
Moments slipping
Each heartbeat close to its end
Tick—tick—tick
Oh, check yourselves in this moment
This heartbeat
This breath
Tick—tick—tick
Oh wicked one, doth riches bring you glory
Doth knowledge bring you happiness
Oh, search out the pearl
Search it out while it may be found
Oh, time slides by
Tick—tick—tick
Hear the clock
Each tick never to be replaced
A heartbeat
A breath

May be the last
Tick—tick—tick

March 1, 2004

Time Spent

How much time do you spend with the Lord
How much time do you spend
Where are your hours spent, My child
Where are they spent
Are they used within the world
Or are they used on Me

March 8, 2004

To Be Blessed

Oh, to be blessed by the Lord
To be blessed with good health
To be blessed with a home
To be blessed with a job
Family
Friends
Country
Oh, to be blessed by the Lord
Is He not good
Is He not gracious
Is He not kind
Is He not forgiving
Oh, to be blessed by the Lord
Dance before Him
Give praises to His Holy name
Oh, bow down before Him
Raise your hands before Him
For He is good

March 6, 2004

To Be Recognized

Oh, we seek our lifetime to be recognized by man
We search him out
That who can raise us up
Raise us up within our world
We strive
We work
We study
So that we may be recognized amongst our fellow man
Oh, when we reach our goal
Are we then satisfied
No
We are not
We continue our
Working
Striving
Studying
Are we satisfied
Have we reached our goal
No
We continue on
A vicious circle
We go on
Where is the end
What really is our goal
Is it man that we seek approval
Is it
Is this really our goal
Oh, look around you
To those that thought they reached their goal
Their utmost
Were they happy, content
As they lay within their coffin, when death takes its bite
Did you see satisfaction upon their faces
Did you see rest upon their death pillow

What did you see
Was there an emptiness
A void
Is this what you seek
Is it
Oh, where really does your heart lie
Is it really with the Lord
Or is it within your world
Oh, you can fool the world
But you can't fool Me
Oh, where your heart lies
There lies your soul
Where does your heart lie
Does it lie within your riches
Your goals
or does it lie within your Lord
Where nothing can touch it
Nothing can compare
Oh, where does your heart lie
Knowledge will cease
Riches will cease
Beauty will cease
Oh, what do you have then
Without your account in Heaven
You have nothing which compares
Which is better
Riches for a time
Or riches for eternity
You decide

March 20, 2004

Too Old

Oh Lord, we are never too old to hear Your voice
Never too old to receive Your blessings
Never too old to come to You
Oh Lord, what plans You have for us
What plans
You saw ahead
Watched the many times we tried walking on our own
Tripping and falling
So many times we tried it on our own
All grown up we thought
We'll walk it alone
We need no help
Yet we trip again
And again
Until we realize we are still toddlers
Still toddlers
Learning to walk
Still need that hand to reach out and pull us up—again
Then we allow the teaching of our parent
Teach us to walk, oh Father
That we may not fall again

March 17, 2004

True Love

Oh Father, how we do not walk in true love to you
Oh, how we say we love You
But how much do we love you
Oh, we plant about our homes words of You
Oh, how we want to show others our love for You
But do we show You our love
Do we enter into a closeness with You
Not to prove to others our love to You
But to prove to You, our love
Oh Lord
Strengthen our love
Remove that which is not of you
Oh, remove that which offences You
Oh Lord, check me through and through
If You find within me anything that is not You
Oh Lord, I pray
Remove that from me
Oh Lord, I desire only You
Only You to take up residence
In this most humble vessel
Oh Lord, only You to reside
Oh Lord, if anything not of you
Oh, check me out
Count me in to be throughly checked
And, oh Lord, find in me only a love for You
Only a desire to be as You want me to be
I am willing to be your vessel
Your vessel
To do as You want
To say what You want
Oh Lord, cleanse me through and through
Cleanse me from toe to head
Oh, check my tongue

For if anything not of You
Oh Lord, check me
Cleanse me
For I only want what You want
Oh Lord, I am but a babe yet within Your realm
But a babe
So much to learn
So much to learn
Oh, I stumble and trip and fall
Yet You pick me up again
Encourage me to stand on my feet
And try again
Oh Lord, how I am yet learning
But, oh Lord
You are my backbone
You are my writer
You Lord, my Deliverer
My Lord
My Master
You, Lord, will guide me
You will handle me as You will
Oh Lord, I have a love for You
That is immeasurable
I love you that neither time nor death can remove
Oh Lord
You alone will control me
You alone will guide me
For Lord
You are my Lord
You, who has brought me from the scum of the earth
Pulled me up
Made me stand upon my feet
You came to me, Lord, and beckoned me
Oh Lord, how You came to me in Your majesty
How You presented Yourself to me
Oh Lord
My Lord

My Lord
For Lord
You are my Lord

April 26, 2004

Turn to You Lord

Oh, when my thoughts turn to you, Oh Lord
Are You not thinking of me at that very moment
Having prompted me to think of You
Oh, what a wonder You are
To tickle my brain to think upon You
For You are thinking of me
Oh, prompt me more often, Lord
So that my thoughts are constantly on You

March 15, 2004

Unrest

Oh, unrest has spread its wings
Upon the people of the earth
Unrest has come upon you
Unrest within the lands
Within the nations of the world
Within the rulers of the lands
Within the church
Within the family
Unrest has spread its wings
It has spread itself upon a peoples
Spread itself
Oh, be ever watchful
I am peace
I am love
Oh, where is the peace
Where is the love
Oh, peoples of the earth
I am peace
I am love
My people are peace
My people are love
Oh, My children
A parent will turn against their child
A child against their parent
Husband against wife
Wife against husband
Oh, My children
Remain in peace
Remain in love
Oh, I have come to separate the truth from the false
The love and the hate
The peace and the war
For you will know My children
By their love

Their peace
These are My children
Oh, remain in love
Remain in peace
My children

April 6, 2004

Upper Room

Oh Father, You have said to wait upon You
Oh Father, how long must I wait
Oh Father, You now tell me to wait in the upper room
I am willing to wait, Father
I am willing to wait in the upper room
I am willing to sit in the upper room, awaiting You
Oh Father, will You come in Your might
Will You come in Your power
Oh Father, prepare me for such a day
Prepare me to hold Your glory
Prepare me to step forth in Your glory
To go out, Father
Go out with Your glory upon me
To go out into the world that any who sees me will know, Father
That You are in me
That they will see Your glory shine out of me
That they will see by the works I do, that You are with me
That many will come to You through Your glory that shines through me
That they will come to You, oh Lord
That Your glory too will go with them
As in the days of the upper room, Father
You sent Your Spirit into that room
The room was filled with Your glory
Filled I say
They walked out with Your power
Your might
They walked with all the boldness
They walked in all authority to cast out demons
They walked in all power to heal the sick
To raise the dead
Oh Father,
Is this what You are about to do with me
Is this what You are about to do with me
Oh Father

Prepare me for such a day
Allow me to settle into the upper room
To await Your filling
Your special filling, Father
Oh, Your glory will come upon me
Your glory will fill me
I shall go out in Your power
Your might
I will cast out demons
I will heal the sick
I will raise the dead
Oh Father, many signs and wonders will follow me
But not just me, Father, but to all who sit in the upper room
They too shall be filled with Your might
Oh Father
You are sending Your rains now
Not just the sprinkle of days gone by, but now the showers are coming
All those who sit in the upper room, will walk in all authority of Your son, Jesus
Christ
Who being raised from the dead, took His place with You, oh Father
He took His place with You.
He promised signs and wonders will follow those that believe
Father, we believe
And now as we await in Your upper room, Your promise is coming
It will fill all those who are sitting in the upper room awaiting, Father
It is not just I that You are speaking to
But You, Father, are speaking to multitudes
Telling them to wait in the upper room
And as You promised, a complete filling is about to come
A complete filling
That we will all go out and testify of You
And multitudes will come into Your kingdom through these days.
Multitudes of sons and daughters
They will come with singing
With dancing
With the timbrels
Oh Father, they will come to You
They will all come to You

None will be lost, none that You gave Your son
None
Oh Father
I await in the upper room
I await Your latter-day rains
Oh Father, prepare me for that day
I await Your pouring upon me Your rains
I await

March 30, 2004

Walls

Oh, what walls have you placed about yourself, My child
What walls have you built
Oh, you have built a wall about you
Solid walls
Oh, brick by brick
You piled them up
Cemented them in you did
You have built them so high
You can barely see over the top of them
Oh, what brought it about
You have carried your pile of bricks
Always ready to go higher if the need be
Oh, is there any wall I cannot break down
Is there any wall I cannot see through
Oh, there is no wall I cannot crumble with My very breath
Oh, you have over the years built your wall
Hiding behind it
Safe and secure
Oh, no walls can exist between you and your Lord
No walls can exist
You must pull down your walls
Pull them down, brick by brick
Come unto Me
Come in where it is safe and secure
Oh, come unto Me
I will care for you
No need for walls to protect
To shelter you from your hurts
For I will fight your battles
I will protect you
I wish for you to have no walls built around you
Pull them down
Come unto Me

May 4, 2004

We Fall

Oh Lord, did You not see the many times we fell
The many temptations that crossed our path
The many times we were tempted and walked away
And the many times we fell into the temptation
Oh Lord, You know our weaknesses
You know when we failed
You know when we walked away
Oh Lord, do not the times we fell out weigh the times we walked away
Oh Lord, how weak Your people are
Oh Lord, how weak we are
We know when we fall
Others do not, they see us as never falling by the way side
But You, Lord, You see
You see
Oh Lord, help us not to deceive You
Help us not to deceive others
Help us not to deceive ourselves
For we are guilty of our weaknesses
But, oh Lord
Help us in our weaknesses
Help us when we fall
Forgive us when we fall
Pull us up lest we fall
Plant a hedge around us to prevent us from falling
Oh Lord, You know the temptations that lie across our path
You know the deceivement that lies awaiting us
You know, Lord
You know
Oh Lord, we are not strong like You
We are a weak people
Weakness born into us for temptation
Oh, how our eyes
Our minds
Our hearts

Our tongues deceive us
Oh Lord, forgive us continually
Keep us from falling into the evil one's path
Oh, keep us lifted up
Cleanse us when we fall
Wash us
Forgive us, oh Lord
For we are Your children, Your weak children

March 19, 2004

We Serve

We serve a risen King
We serve the Lord
He is the Almighty
He reigns over all
He is our Lord
Who can compare
Who can stand in unity with Him
There is no one
There is nothing that can compare
For He alone is Lord
Our God Almighty
Is He not the One that moves the winds
Do not the trees look up
Do not the stars cry out
This is our Lord
Our Master
Our King
Our Ruler
Our Father
Our Friend
Who can compare
There is none
Only the Almighty
Search yourselves, oh ye earth
Search yourselves I say
Where does your heart lie
What are your loves—your true loves
Search yourselves I say
For I see your hidden sins
I see your thoughts
Drop them I say
Come to Me pure
Let go of your evil ways
Then—I will bring about wonders

You will see my miracles
There has been none like them before
You will stand in awe
Then the people will see
Their cries will reach the Heavens
I will bring them to their knees
And they will know
That it is I
I am the One
I am the Almighty
I am their King

February 11, 2004

What Do You See

Oh, what do you see when you look upon another human
What do you see
Do you see what is hidden under the flesh
What types of spirits lie hiding here and there within one's body
For we all have them
Oh, when someone angers you—is it that person who angered you
Or
What lies within them
Oh, when someone lies to you is it that person
Or
What lies within them
What of all the other things that occur that offend, hurt, rob, murder
Is it the person that has done it
Or
What lies within them
Oh, we all carry within ourselves spirits both good and bad
For we all sin
All
What has made its residence within you
What types of spirits live in you
Are there more evil ones than good ones living in you
Oh, start to clean out your house
Cast out what is evil within you
Replace that void with good
Or
The evil will return seven fold
Clean your house

March 1, 2004

What Do You Want

Oh, what is it you seek after, My child
What do you want of Me
Have I not said to ask of Me
And I will grant you your desires
From the riches of My Father
Have I not told you so
Oh, ask of Me
What do you seek from Me
Have I not already given to you
Have I not already given to you
Oh, all things are done
All things finished
I have already granted you
Your desires, your wishes
Have I not accomplished all that My Father willed
Have I not
I went about My Father's will
I went about My Father's will
It is finished
Oh, is it sickness that you ask of Me to leave
Is it that ye are blind, deaf, lame
Oh, what is it you ask of Me
Have I not already granted you your desires
I went about My Father's business
I accomplished all He asked
Yet, ye still seek from Me
Oh, ye must look beyond the impossible
Look beyond your unbelief
See the end result
See it
It lies just beyond you
It is there
Answered
Open your eyes

Open them
For what ye ask of Me
I have given to you
As I went about My Father's business
And accomplished all His will
So shall ye too shall go about all
As I have done My Father's will
So ye too know My Father's will
That you ask believing without doubt
That He will
That He will given you the desires of your heart
What are your desires
To be healed
You are healed
Is there anything too difficult for Me
Is there anything too difficult for Me
Oh, what is impossible for man
Is possible with Me
Oh, nothing is impossible
Oh, have I not created all
Have I not gathered the dust of the earth
And molded man from it
Have I not sewn man together
Yet you think it is impossible for Me
To heal the lame, the blind, the diseased
Oh, where is your faith
It lies just beyond you
Step out
Reach for it
Oh, nothing is impossible for Me
Ye have asked
Have I not accomplished all My Father's will
Oh, His will is you live life abundantly
That you life be full
Oh, ye have faith to know that your sins have been forgiven
Ye have faith to know that
Oh, where is your faith to see beyond
Have I not said, ask, and I shall give you according to My Father's will

Oh, His will
Oh, see what His will is
For I did My Father's will
Oh, see His will for you
Ye are healed, My child
Ye are healed

April 23, 2004

What Evil

Oh, what all has man imagined within his heart
What has he imagined and brought forth into form
Oh, what has man done with these forms
Has he not sent them out into the airwaves
These forms coming and going
Oh, what amount of forms there are
Going and coming
Crowding the skies they do
Crowding them
Oh, how evil man is
How evil
Is there nothing you have not imagined within your heart
And brought forth into form
Nothing at all
Oh, how the air waves are filled
So filled
Sent throughout the earth they are
Sent across the waters
Sent across the lands
Language has no barrier
No barrier at all
Oh, how evil man has become
How evil
For within man hides the greatest of dirt and filth
The greatest amount
Oh, you evil one
What evil lies with in you
What evil
To fill My skies with your filth
To fill them with your evil thoughts
I will squash them from you
I will squash you, oh evil one

March 15, 2004

What Lurks

Oh, what lurks behind you, My child
What hides itself in the darkness
Awaiting
Awaiting
Awaiting for you to trip
Awaiting you to fall
Oh, he is there—well hidden
Oh, don't let your guard down
Don't let it down
For he lies in wait
Awaiting for a weakness to overcome you
Then he will trip you and snare you
Oh, be well guarded
Oh, watch your back
Always
Don't let a moment go by without watching your back
Keep your guard up
Never let it slide
Keep covered
Walk with boldness
Never allowing your guard to slide
Keep yourself covered
Be prepared
Keep your sword handy
Never let your guard down
For he who lingers in the darkness desires you
Will stop at nothing to try and snare you
Oh, keep your protective gear on
Keep it on

March 13, 2004

What Riches

What riches we desire
The ability to purchase our desires
No worries sit upon our doorstep
Oh, what riches can buy
Our dreams fulfilled
The home of our dreams
Travel, to be able to go anywhere we wish
Just to have the riches of the richest
Oh, what a desire
And what do we do once we have it all
And done it all
Do we stop buying
Do we stop buying
Do we stop travelling
What then do we do
With nothing left to buy
No where we have not been
A void appears
A void
Nothing left to desire
Then what
What then have we gained
What then have we lost

March 10, 2004

What Sins

Oh, what sins I was born with
Having being passed down from my parents
Then as I travelled through each day, each minute
How many more did I pick up and store away within myself
Daily picking, and packing away
Oh, what weight one carries around
Picking and packing in
We walk around with this heavy weight of sins
Why do we do that
Is it necessary to load up and carry it with us daily
Have not our sins been removed
Have they not been taken away
Yet, here we are, walking bent over
With the weight of it all
Pitying ourselves for all our problems
Oh, foolish one
When you can step forth
Weight free
Get rid of your baggage
Get rid of it
Toss it aside
Walk lightly
Oh, foolish one

March 17, 2004

Where Can You Run

Oh, where can you run from the Lord
Can you hide within your dwelling place
Can you hide under the cliffs
Can you hide in the depth of the sea
Oh, where can you run that I cannot see
Oh, why are you running
Do you have something to hide
Something that causes you to run
Come out in the open
Pour out your sins
Forgiveness is there
Don't hold it in
It is poison to your system
Sickness and diseases arise from it
Oh, search yourself where your sins lie
Bring them out in the open
Allow the light to cleanse it
Oh, cast it out
Your thoughts are not hidden from me
I look upon them awaiting your move
Which way would you go
Will you enter into the sin that was craftily deposited into your mind
Or will you turn your back
Walk away
What is your decision
Oh, cast it out
Stop
Search
Cast out
Remain

March 6, 2004

Where Would One Look

Where would one look for my Lord
Are You to be found within the winds that blow
Are You to be found amidst the stars
Are You in the depth of the seas
Can the birds tell of Your presence
Oh, where would one go to find my Lord
Where would one look for my Lord
Can the birds tell of Your whereabouts
Can the trees point in Your direction
Oh, where can one find you, my Lord
You are so hidden within the heavens are You not
I listen for Your voice within the winds
I listen for You in the music of the birds
I listen for You in the seas that roar
Oh, where can one hear Your voice
Oh, how majestic You are to hide within one's heart
Your voice I hear from within myself
Oh, how wonderful You are
Yes, the birds sing their songs to You
Yes, the winds carry You
Yes, the seas are about You
Oh, how huge You are
The wonder of You surround me
Oh, Lord, my God
Your presence is everywhere
Is it not

March 1, 2004

Who Can Measure

Oh, who amongst you can measure My ways
Who can step forth and claim that it's him
What measure would you use
Oh, can man see beyond themselves
Can man look within the heavens
What a child you are
You are still within your growing years
There is much to learn
So little time
Has man cast down his all
Step forth, ye one who can measure My ways
Step forth
Is there one that can read My ways
Is there one that can read My thoughts
Oh, the foolishness of man
Because you fly in the heavens
I can cause anything to fly in the heavens
Does man believe this is a first
There is nothing new that has not been
Nothing
I will bring about things new
I will create things new
Simple man cannot comprehend what I am about to bring to pass
Oh, how utterly simple man is
I look upon you as you struggle to gain all
Know all
Be all
Ha
You spend your entire life gaining
But what riches will you take to your grave
What can you take that is yours
Can the grave hold your riches for you
Where will you use your knowledge in the grave
Your riches

What can money buy there
Yet your entire life—a lifetime is used on what your gains might be
The riches are mine to give
The knowledge mine to give
Your life mine to give
I can remove your riches
I can reduce your intelligence
I can take your life
What right does a child to tell their parent that their knowledge was of their
own
Their riches acquired on their own
That they control their life
Woe to that child
Bow before your parent
Give thanks for what you have
You have what you have because you are blessed
Oh, My child
Open your eyes
Open to My words
Open your heart
There is so much I have planned for you
Use what I have given you
Stretch it
It will be measureless, My child

February 29, 2004

Why Do You Search

Why do you search
You run here and there
You look for health
You look for love
You turn this way and that
Why do you not look this way
You look all over except My way
Why, can you not see
Can you not just once glance this way
I await your glance—just one glance
My heart pants after you
You hurt without even knowing
You seek this way—that way
Yet there is no satisfaction
Can I not fulfill that love
Can I not fulfill that health
I have it all
I have told you so
I have it all
You reach your end
Yet do you cry out
Must I drop an opener for you to see
I will not do this for long
I will not await much longer
Time is running out
You've had your chance to seek me
You've had your chance to have perfect health
Yet you sought other means
There are no other means
I created all
I created you
Can I not renew your strength
Can I not give you love—perfect love
Can I not give you perfect love

I can do all
Yet your faith is small
Your faith is in man
I created you
I will not stand in the background
You cannot hold me back
I will show my might
I will show my mercy
I will show my love
You just look this way
Just a glance, I will see
You are my love
You are my child
I love as no one can
I love as nothing else can give
Give me that chance
I will show you, my child
A love like no other
Health like no other
A oneness that nothing else can fulfill
I will bring you in,
I will encampeth round about you
I will bring you into a shelter like no other
There is no-one that can fulfill what I can give
No other

January 30, 2004

Wickedness

Doth wickedness lie in your path
Has it crossed your path
Have you ventured its way
Did you find all you wanted
Did the excitement suffice your desires
What are your desires
Do they quench your thirst when taken
Did you still need more to quench your thirst
What other desires are now needed
What more shall be fed to quench your thirst
Is there an end
What more is needed
Will this last desire be the last—or will you need more
Is there an end
Oh, think about what more you desire
Will that be the end of it all or will there be another
There is another way, my son
A desire that will continue to fulfill your desires
Your needs
Stop—
Try this way
See if your thirst is not quenched
All that you desire fulfilled
Seek this
Try it and see

March 5, 2004

Wisdom

Oh, where can one obtain wisdom
Is it to be found within your books
Can the wise direct you to find it
Can you travel to the east and obtain it
Is a person born with it
Can it be bought
Oh, to have the wisdom of Solomon
Oh, where can this wisdom be found
Oh, ask for wisdom
Ask I say

March 6, 2004

Wise Man

You see the sun come up
You know day is imminent
You see the sun go down
You know night time cometh
Yet you see the times you live in
And cannot see the storm approaching
A wise man knows when the sun is dawning
He knows when the night is coming
A wise man knows of the approaching storm
And prepares his house

March 5, 2004

Wives

Oh, how many of you provoke your man
Provoke him to anger
Provoke him to fall into sin
Oh, how you anger him
Anger him with your ever nagging
Your nagging is sending him out
Out to fall into sin
Oh, you drive him to sin
Oh, helpmate to your man
Where is your place
In the household
You nag and nag
Lower your man
In the eyes of your children
Lower them within the family
Oh, you take away all his manhood
Strip him of where he should stand
Lower him beneath the children
Down to the floor with the dogs he crawls
You, you wives, placed here for your man
Placed here to be his helpmate
You, part of his body, one you are
Yet, you take that part of the one
Separate it
Dwindle it down
Raise yourself high
Oh, woman, I speak to you
Step down from your high and mighty place
Step down
Stand side by side with your man
Be the helpmate you were intended to be
Stand by him
Lift him up to where he belongs
Lift him up to be the head of his house

Lift him up in your house
And I will bless this house
I will come and bless this house
I will bless this man and his household

May 12, 2004

Wondrous Works

What wondrous works You are doing with Your people
Oh, what wondrous works
How Your hand moves amongst us
How Your very breath fills us
How Your very being powers us
Oh, Lord, how magnificent You are
How wonderful Your works are
Oh, how You chose us
Us, Your humble, sinful people
How You pulled us up
Pulled us up, You did
Made us stand upon the rock
You Lord, the rock
Oh, how death cannot destroy us
Oh, how pain cannot destroy us
For You, oh Lord
Our deliverer
Our master
Our teacher
Our Savior
Our Lord
Will bring all things good about
You, oh Lord, our Savior
Our deliverer
Will bring about a new thing
Oh, Lord, You will bring about a mystery
A mystery
Oh, Lord, how I bow down to You
You, My Lord
My Master
My Lord
My God
My deliverer
My Savior

Oh You, Oh Lord
Your very presence
Your very presence
Is so powerful
So very powerful
Oh, can one stand before You
Can one accept all You have
Oh, Lord
How wonderful You are
How magnificent You are
You, who made the heavens and earth
You, who created all things
For You, oh Lord
We were created for You
Oh, Lord, let us be so humble
Humble to our creator
Oh Lord
You, You alone
You Lord
You alone
Have made a way
Have made a way
For us sinful ones
Us sinful ones
Oh, how we all fell
We fell before You
Oh Lord
Oh, how wicked we were
Oh, Lord, You alone have made a way
Oh, Lord, us sinful ones
Us, sinful ones
Oh, how wicked we became
But oh Father, You saw down the channel of time
Did you not
You saw all would be lost
All you created, us
Oh, Father, You made a way
No more sacrifices

No more sacrifices of innocent lambs
For men to continue in their sinful ways
Oh, Lord, You sent out your messengers
You sent out your prophets
You sent out Your kings
Oh Lord
All was lost
For man continued in their sin
Their sin until the next sacrifice
But Lord, You saw down the time channel
You saw man and all our faults
All are guilty before You, oh Lord
All are guilty
From Adam until the end, all are guilty
And Lord, You saw
The sacrifice to be made
And Lord, You provided a sacrifice
Lord, You provided
Your only Son
Your only Son
Your only Son
Oh, Lord Jesus
The son of God
Oh, how You too saw our wickedness
Our wickedness
You saw
You stepped in
You stepped in
Took the place of all the slaughter of lambs
You, Lord, took their place
You, stepped in
Oh Lord, what glory
You have brought
What power You brought
What forgiveness You brought
For we are saved
By the blood of the lamb
Bought with that blood

Once and for all
No yearly slaughter
No yearly taking of the lamb
For all is paid
All is paid
Oh, how the people of God rejoice
Rejoice ye people of God
Rejoice
For our sins now paid
Our sins now removed
No daily, weekly, yearly slaughter
All have been paid
Paid by the blood of the Lamb
Oh, merciful
All is paid
Oh, how the people of God rejoice
Rejoice ye people of God
Rejoice
For our sins now paid
Our sins now removed
Oh, merciful Lord
You, who stepped down
You stepped down from glory
You, my Lord
Savior of all
Deliverer of all
Forgiving of our sins
Bringing us in
You, oh Lord
Have brought us in
You are bringing us all in, oh Lord
Oh Lord
You have brought us all in
All Your children
All

April 20, 2004

Words Go Forth

Oh, what happens when My word goes forth
And is sent amongst the people
And
What happens when the evil one sends forth his word
Oh
That word sits in wait
Watching, watching, watching
Oh, there is one without a garment
Oh, enter in, enter in
Another sits and waits
Oh, another with its garment ripped and cracked
Oh, enter in, enter in
Oh, here is another, the garment all worn, not yet hole riddled
Oh, hitch a ride, hitch a ride
Oh, enter in
Oh, leave alone
Oh, hitch a ride
Oh, be wary, My children
Don't walk around without your garment
Don't have a garment that is torn and has not been mended
Don't walk around with your garment so worn, and has not been looked at for
years
Oh, keep your garment in top-notch condition
Always check it over—daily
Keep the repairs of your garment up
Wear only the freshest of garment
For the evil one has sent forth his word
And awaits for the passerby
Watching each and every one
Be warned, keep your garments on and checked

March 20, 2004

Words

Oh, words are mere nothing, Father
Without Your power
Oh, Father, Your words
Your words, Father, cut deep
Hit the source
For which it is meant
Oh, Father, we all are guilty of sin
All are guilty
All feel the piercing of Your word
All feel the cut of the sword
Oh, the two sides of the sword
Cuts down both sides as it penetrates
Oh, Father, your words
You give power to Your words
Oh, I can do nothing, Father
Nothing without You
I can do nothing
Oh Father
Your power
Your might
Oh Father, continue to use this vessel
Oh, I am but nothing
Without You
I am nothing
A mere nothing
But Father, with You
I am something
Used of You
I am something
Oh Father, continue to fill me with Your words
Your words
Continue, Father, to fill me
I pray

May 5, 2004

Worship

Oh, how easy it is to worship the Lord when the sun is shining
Or when the stars glisten in the heavens and the skies so clear
When all things are in order for you
Good health
Good job
Nice family
But what happens when the sun is not shining
When the skies are overcast that the stars are hidden
When nothing seems to be in order in your house
Poor health
Unemployed
An uprising within your household
How easy is it to worship the Lord then
How easy is it
Has He changed from when things were right for you
Does He not cast the sun upon the good and evil one
Does He not show His stars to the good and evil one
Does He not bestow all things in order for the good and the evil one
Oh, worship the Lord in good times
Worship Him in bad times
For He is worthy to be worshiped

March 13, 2004

You Have Walked

This is given to those that have been put "on hold"

You have walked before Me
You have been tested
You have been tried
You were not put upon the shelf to collect dust
But you were put in a safe place
A place of holding
A place where I kept My eye upon you
Watched you prove yourself
Now, I am about to place you out in the open
You will now be filled
You will be filled not like you were
But filled
Overflowing
Be prepared, My child
For you will have an overfilling that will pour out and over you
The overfilling will pour out and over you
Where you will take it out onto the streets
Where your flowing will touch those that want to be touched
They too in turn will touch others with this flowing
Can you handle what is going to come upon you, My child
Oh, My obedient and faithful servant
I am now about to bless you
Bless you with My power and My might, My child

March 16, 2004

Your Children

Oh, what beautiful children You have, oh God
What beauty they have
Oh, beauty beyond any imagination
A beauty anyone would desire
Oh, how they stand out in the world, oh Lord
Their beauty out stands any other
Oh Lord, Your children
Is it not so that beauty comes from the parent
And oh Father, such beauty your children have
Oh, how beautiful You are

March 13, 2004

Your Days

The days of your life are numbered
How many days have passed on by
How many days remain
Your days come and go
What have you done with the days you have used
What will you do with the days that remain
Each day—a gift
What waste have you done with your days
What good thing have you done
Does the scale tip more one way than the other—
or does it sit in a balanced state
You do not know of the days that remain
What are your plans for them
Will the scales continue to lean in one direction
Or will it stay centered
Idle
Motionless

March 5, 2004

Your Earnings

Oh, foolish one
You put aside your earnings
For your future
For a home
You put aside your earnings
For your child
For their learning
Yet, you know not the time allotted you
You know not the days given you
Yet, you put aside your earnings
For tomorrow
You know not what tomorrow brings
You know not what your child will want
Yet, you put aside
Put aside
You cast your eye on future years
Yet you know not how many years I have numbered for you
You know not
Yet you place your earnings aside for those days
Days not yet come
Days not yet come
You worry about days to come
Days that are not here
May not come
For you
Yet, you place your trust in your earnings
Your savings
Your future
You have no earnings
No savings
No future
That does not belong to Me
They are Mine, and I grant to whom I wish
I have molded you

I am the potter
You are the clay
Who can tell the potter how long they shall live
What they will be
Yet, you, the clay
Tell the potter, I, the potter
How you shall be made
How long you shall last before crumbling
What your offspring will become
I, am the Potter
I will decide
You are My children
My children
I have for My own
I created you for My glory
Yet you—cast Me aside
Do not consult Me for your future
Your life
The life of your offspring
Have you placed your life in My hands
Have you placed your children into My hands
Oh, cast yourself upon Me
Cast your seed upon Me
Cast your riches upon Me
I will guide
I will orchestrate their lives
I will bring about a new thing with them
Oh, cast your cares upon me
For I care
I care
I care

April 3, 2004

Your Feet

Oh, where do your feet take you, young man
Do they take you into the house of sin
Do they take you to places unknown
What secrets are hidden within your feet
Does one know where they tread
Does your loved one know from where they left
Oh, watch yourself, young man
Watch yourself—for nothing is hidden
Nothing
Be sure you will be found out
Oh, cleanse your feet
Turn yourself around
Take your feet in the opposite direction, young man

March 5, 2004

Your Hand

Oh Lord, how You draw Your hand across the skies
Diminishing daylight
Bringing forth night
Oh, how the sun slowly moves downward
Bringing with it a mass of light
Within the approaching night
Oh Lord, how when You come
Oh, such a massive light
Breaking forth into darkness
Oh, how all eyes are drawn to the light
And darkness seems to disappear
Oh Lord, how You made the heavens and earth
How You chose where and when the sun and moon would go
Oh Lord, You planned the sun to give us warmth
To bring forth food
Oh, how You judged the moon on how it would move the tides
Oh Lord, You are the majestic of all
You alone have brought all things about
Oh, how You paint the skies
Each day
Each night
A different scene
Oh, such a painting You create
Oh Father, how You have moved within our lives
All our lives
Moving all obstacles from our way
To bring us into You
Oh, how You have played the game of chess
Each move
A change for us
You have had a hidden meaning to why it is occurring to us
Oh, such devastation for us at the time
But Lord, You see the end results
You move the right people across our path

To have such an impact on our lives
The scene changes
And changes
And changes
Then Lord, the end result
A blooming flower arises
Oh, how You saw the end result
Me, from the bottom of the heap
Lifted me up step by step
Error by error
Hurt by hurt
You, Lord, did always have Your hand on me, didn't You
You allowed me to fall
Bruise myself
Get up and start again
Oh Lord, You stood back while I hurt myself badly at times
Anger slipped from me
Anger at my loved ones
But anger at You
Yet Lord, You watched me
Awaited Your time
You are so ever patient, my Lord
Ever patient
How You saw the end
While I muddled in my grief
Oh Lord, what lessons I have learnt
What lessons You have taught me
Oh, lessons I can now teach to others
Roads I walked on
Difficulties I endured
Oh Lord, you have proved a way for me
But You have also proved a way for others through me
Oh Lord, You are My great King
You are my Lord
Oh Lord
I see the day when we see You
I see the day when we sit down and dine with You
Oh Lord, what You have planned for us

Your children
Oh Lord, soon we will all be home
Oh Lord, how the music will flow
Oh Lord, eternity with you
Eternity with the praises
I await that, my Lord
But not before You have finished my job here
Not until this vessel is useless to You
For Father
You are my Lord
My Lord now
My Lord forever

April 27, 2004

Your Hearts

Oh, where do your hearts lie, you inhabitants of the earth
Do they lie in want of Me
Search yourselves
I say search yourselves
Don't fool yourself
There are none that can say I sin not
No, not one
Search yourself—cast out that which pulls you down
I say cast that out
Look around you
Do you see purity anywhere
Do you see a sinless nation
The sins of My people disgust Me
Turn yourselves around
I say turn yourselves around
There is not one that can say I am pure
No, not one

March 8, 2004

Your House

Oh, what we hide within our house
What weaknesses
What evil
Do we hide within the walls of our house
Hidden from view
Hidden from the eye that peers at us
Oh, how hidden our faults are
How hidden
Hidden from all who look upon us
Hidden
Oh how we hide within the walls of our house
Where no one may enter
Locked up tight
We are
Within our house
Hidden are our evil ways
Our faults
Our sins
Oh, so hidden from the public
We are within our house
Oh, do you not know your body is your temple
The temple of the Lord
Oh, what have you brought into your temple
Your house
Oh, what have you brought in and hid from the public
Oh, what have you brought into your temple
Oh, your temple
Where you have so cleverly hidden your sins
Your evil ways
Oh, clean your house
Brush out the cobwebs that are so hidden where evil is tucked away for a rainy
day
Oh, brush out your house
Throw out what does not belong

Don't hold onto something that will bring about decay
Oh, get rid of that which will fester
Oh, wash the walls clean
Wash the walls clean
Sweep out your house
Clean it clean
Open the windows
Open the doors
Allow a fresh breeze to flow through
Bringing with it the sunshine
Bringing with it healing
Peace
Rest
Oh, My child
Present your temple holy unto Me
Present it before Me
Clean and pure
Present it before Me
I say

April 3, 2004

Your Life

What have you done with your life, My child
What have you done
What have you done throughout your life
What certificates can be hung
What diplomas to be shown
Where in My kingdom can they be hung
Oh, what have you done with your life, My child
What gifts from Me have you used, My child
What gifts from Me are shown
Oh, where through your life have I shone through
Oh, where
Are not the gifts implanted in you
A gift of the Lord
But where have you been thankful of my Lord
What lies within your future years
What lies ahead
Does your gift abound
Tread carefully, My child
Tread carefully
Gifts may be given
A thankful heart retains
Be careful, My child
For your gift may be removed

March 8, 2004

Your Name

Oh Father, I don't ever want to shame You
Or disappoint You in any way
Oh, help me I pray
I don't ever want to see Your face as I shame You
Or disappoint You
A Father looking upon His child as they fall into shame
Oh, how hurt a father is
His child
His child
Brought shame upon themselves and upon their Father's name
Oh, protect your name
Honor it
For a Father is proud of His name and you carry that name, young one
You carry the name of your Father
Who is your Father, young one
What is your last name
Whose name do you travel with

March 11, 2004

Your Robes

Oh, your robes are being prepared for you, My child
They are being sewn with the finest of threads
Not man-made fibers
But threads of
Truth
Purity
Honesty
These threads have being woven into the finest of cloth
White as snow
White as snow
They have been cut with the finest of scissors
Cut with the precision of truth
Carefully sewn with the finest of machines
Honesty
A robe made of purity
White as snow
White as snow
Your name is sewn into the very threads
Your name
Not your earthly name
But the name I have given you before the foundation of the earth
Your robe is hung out on the line of honesty
Awaiting you
Awaiting you
Your robe
White as snow
Your name engraved within the threads
Your name, My child

March 16, 2004

Your Spirit

Oh, how Your Spirit moves
Gently coming
Oh, how we get washed inside and out
Oh, how Your Spirit moves
Oh, how gentle You are
Yet so powerful
Cleansing me, moment by moment
Oh, daily cleansing me
Oh, daily renewing me
Oh, Holy Spirit
How You come in Your power
How You move upon me
Oh, how changed I am
Oh, how cleansed I am
How thankful I am that You drew me
Me, You chose
Me
Oh, how You came and drew me in
Drew me to, my Lord
Oh, how You drew me
Oh, how thankful I am
Oh Holy Spirit, continue to always draw me in
Draw me in yet further
Oh, Holy Spirit
Daily cleanse me
Daily, I ask

March 28, 2004

Your Thoughts

Oh, what thoughts run within your mind
Are they thoughts of
Lust
Greed
Envy
Jealousy
Fornication
Homosexuality
Deceitfulness
Lying
Are any of these thoughts lingering within your mind
Your thoughts run into a blueprint for one to build on
What type of a structure would be conceived
Oh, change the plans
Oh, empty out your thoughts
Replace the thoughts
See what type of structure could be built
A glorious structure
A firm foundation
An eternal house

March 5, 2004

Printed in the United States
30381LVS00003B/109-138